D1871695

Disaster Vulnerability Management and Mitigation

Disaster Vulnerability Management and Mitigation

By

Dr SEEMIN MUSHIR

KALPAZ

©Author
All rights reserved. No part of this work may be reproduced, stored, adapted, or transmitted in any form or by any means, electronic, mechanical, photocopying, micro-filming recording or otherwise, or translated in any language, without the prior written permission of the copyright owner and the publisher. The book is sold subject to the condition that it shall not, by way of trade or otherwise, be lent, resold, hired out, or otherwise circulated without the prior publisher's written consent in any form of binding or cover other than that in which it is published.

The views and opinions expressed in this book are author(s) own and the facts reported by them have been verified to the extent possible, and the publishers are not in any way liable for the same.

ISBN: 9789353240387

Published, 2019

Published by

Kalpaz Publications
C-30, Satyawati Nagar,
Delhi – 110052
E-mail: kalpaz@hotmail.com
Website: kalpazpublications.com
Ph.: 011-47034999, 9212142040

Printed at: G. Print Process, Delhi.

Dedication

DEDICATED TO MY PARENTS,
MY TEACHERS AND STUDENTS

Contents

SECTION - 1

The Disaster

Section - 2

Disasters in India

List of Tables

List of Figures

List of Maps

Foreword

Disasters are occurring with increasing frequency now. Disasters have always affected mankind and will continue as long as life exists on the Planet. These occur all over the world. India is one of the most vulnerable regions of the world in respect to disasters.

The emerging scenario is indicative of a vulnerable world including India. The disasters cannot be stopped but their effects can be reduced up to some extent by careful planning and preparedness.

This book defines disasters that are prevalent in the world and the ways adopted for their management and mitigation. Risk and vulnerabilities are viewed here as a manifestation of marginalization of a large section of society, their vulnerable living conditions and lack of access to basic services.The vulnerability varies between developed and developing regions of the world.The former group links vulnerability to factors defining the standard of living, state of infrastructure and access to resources. For the latter group it is the issue of survival. Meaning of vulnerability changes within each group of people based on the cultural contrasts worldwide.

The adverse impacts of physical phenomena like earthquakes, landslides, floods, droughts and cyclones are observed on environment and human settlements, particularly on the people living in the seismic zones, floodplains and coastal belts.

The present work is an effort to explore and investigate disasters, their typology, occurrence and management. The book is designed for the students of disaster management and for those who are preparing for various competitive exams including civil services. This book has recent data and maps presenting current scenario of the problem in the world and India. The book has recent information so that it can be helpful to planners and disaster managers as well as students of related fields.It explains the occurrence of all the disasters and the possible ways of their mitigation.

Majid Husain

Professor and former Head

Department of Geography

JamiaMilliaIslamia, New Delhi

Preface

Natural disasters are a devastating, but undeniably inevitable part of life and society. Because of this fact, many of us tend to believe that they happen as an act of nature, purely out of the human control. We need to start realizing that this is far from the truth and it is this attitude that is stopping us from learning from our mistakes. The contemporary world inaccurately labels various disasters as 'natural' when in fact, when looked at more closely; there is an enormous correlation between 'nature' and 'society'. There are many flaws in the human preparation for natural disasters that have equated in death and injury where it could have been minimized. Although the occurrence of a disaster can happen at any time, Natural disasters, such as earthquakes, hurricanes and cyclones, floods, and tornadoes are some of the traumatic experiences that may cause posttraumatic stress disorders. This type of disorders and other disaster related psychological problems are prevalent after natural disasters. As a result it is important to analyse those individuals that develop psychological problems especially to prepare and plan interventions both in the short and long term after disaster. Earthquakes have quickly become one of the most highlighted natural disasters. Unlike other natural disasters, earthquakes occur without warning.

These natural disasters are exacerbated by human cause such as weak infrastructure, faulty political system and a complacent civil society. Nature is blamed as enemy, and true responsibility is not accounted.

Our Earth has suffered a great deal from reoccurring natural disasters that have repeatedly put a strain on people's lives. A natural disaster is a sudden event, an accident or a natural havoc, that causes great extents of damage or multiple deaths. Over these past years a numerous amount of these disasters has been seen happening all around the world. Tragedies of towns getting torn apart were reported as well as the occurrence of many deaths, disabilities and shelter damage. People have been losing relatives, friends, shelter and property. Even though these disasters cannot be stopped from occurring, recognizing their danger and their effects is essential.

Humanitarian aid, treats the symptoms rather than the root cause of catastrophe. The nature cannot be predicted correctly and fully controlled it is the preparedness which can reduce the impacts up to some extent.

It is a great experience to pen down this book, which help too many students in India and abroad of geography and disaster management.

Acknowledgement

I am grateful to the NDMA (National Disaster Management Authority of India) for providing data, maps and other relevant information used in this study. I also wants to thanks to IMD,

(Indian Meteorological Department), NIDM (National Institute of Disaster Management) for their extended help in the completion of this book.

I also wish to put on record my obligation to the librarians of Dr. Zakir Hussain library, and Delhi University Library for their timely help in providing me relevant literature and data.

I have special feeling of gratitude for Prof. Majid Hussain a renowned geographer, my teacher and mentor for his guidance and support in the completion of this book.

I want to pay special thanks to Prof. Masood Ahsan, The head of the Department of Geography Jamia Millia Islamia, New Delhi, my alma mater and working place where I am working from last fourteen years, for his outstanding support and guidance.

The teachers and the non-teaching staff members of the department have been the backbone of my whole educational and professional carrier, they all are very kind and supportive for me specially my all the teachers. Prof. Haseena Hashia and Prof. Mary Tahir deserve special thanks for their help and positive suggestions in the completion of this work.

Finally I must acknowledge my gratitude to my family for their motivation as well as moral and financial support and encouragement which have enabled me to complete this work in time.

Seemin Mushir

SECTION - 1
The Disaster

Chapter One

Disaster

A disaster is a condition of emergency which causes serious disruptions to a society. Disaster management is the set of counter measures which are designed to counter disasters and their aftermaths.

A disaster may be defined as it may be "an event, which threatens a society with major unwanted consequences as a result of the collapse of precautions which had hitherto been culturally accepted as adequate." (Turner, 1976).

Vulnerability + Hazard = Disaster

A disaster occurs when a system's resilience capacity is lost.

A disaster is catastrophe which completely obstructs the normal ways of life of people of a region and brings heavy burden on human society by causing losses of various types like economical, and environmental, which the community cannot bear and become a disaster victim. Earthquakes, tsunamis and landslides, floods, Avalanches etc. are called as disasters when they cause heavy destruction of life and property. Hence, when there is a tremendous loss to property and human population in an area that is thickly populated it is called a disaster.

There are man made factors that help in turning a hazard into a disaster. The anthropogenic activities such as deforestation, emission of greenhouse gases, vehicular discharge are changing the environmental chemistry thereby problems of change in climate and global warming are arising and due to their influence weather related hazards in many parts of the world have resulted in an increased frequency of floods, cyclones, landslides, earthquakes etc. That leads to widespread destruction. Anthropogenic activities like burning of fossil fuel, emission of greenhouse gases and deforestation are the root cause of increasing frequency of natural disasters

Hazard

Hazard is a danger to life, health, environment or property. Natural hazards are earthquakes, floods, Tsunami, forest fires, landslides, and droughts, volcanic eruptions etc. They cause a lot of destruction and are natural phenomenons that occur without any prior warnings. When any of these hazards takes place in an area that is uninhabited, it causes no harm to human lives or property, now it can be said that a hazard is an event that has potential to cause widespread destruction and loss of lives and property. But, when a hazard strikes an area that has no human population, though it still has destructive properties, it is not termed as a disaster.

There are several types of hazards, such as Physical, like earthquakes and landslides, leakages of chemicals like Bhopal gas tragedy, Biological like spread of epidemics, Psychological, and Radiation Hazards.

Hazards and disasters are both natural as well as man induced.

Table 1: Difference between Hazard and Disaster

Hazard	*Disaster*
A hazard is that condition where there is a threat to life, health, environment and property.	A disaster is an event that completely disturbs the normal way of life of people of a place. It brings human, economic, and environmental burden which the community cannot bear.
Hazards are an integral part of our environment and cannot be prevented. They pose threat to life and property.	These hazards are known as disasters, if they cause heavy destruction of human lives and property. When a hazard becomes active and is no longer just a threat, it becomes a disaster.
A hazard is less dangerous if we took preventive steps immediately when it discovered.	A disaster is an emergency which strikes unannounced and personal mitigative steps are required to save one's own self when it is happening.

Types of Disasters:

1. Natural Disasters
2. Man Made Disaster

Natural Disasters

Disasters by Continents 2005-2014

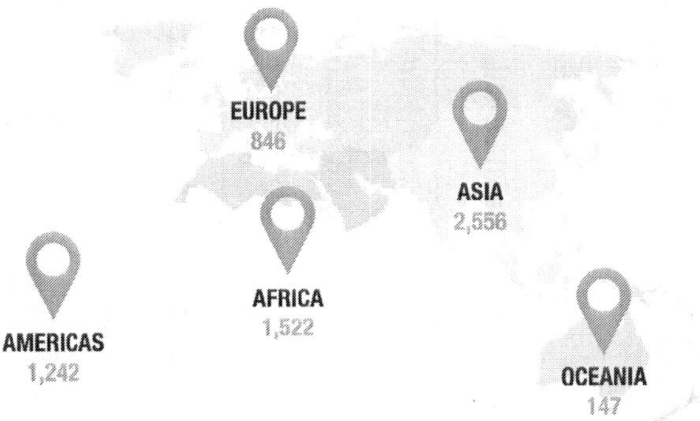

Source: world Disaster report 2016

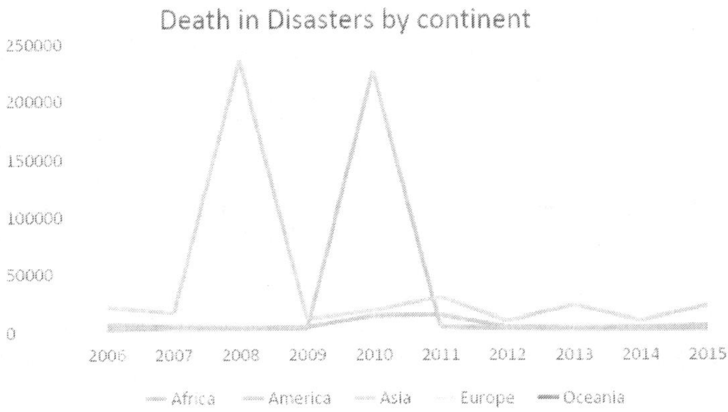

Fig: 1. based on data provided by EM-DAT, Cried, University of Louvain, Belgium.

Fig made by the author

Table 2: Total Number of Disaster by continent (2006- 2015)

Continent	*2006*	*2007*	*2008*	*2009*	*2010*	*2011*	*2012*	*2013*	*2014*	*2015*	*Total*
Africa	202	184	173	156	135	165	124	115	101	116	1471
America	105	133	143	115	146	131	114	106	116	124	1233
Asia	308	262	240	233	253	235	210	228	228	240	2437
Europe	98	104	58	75	99	49	91	69	80	70	793
Ocenea	18	11	13	19	18	15	14	12	12	24	156

Source: EM-DAT, Cried, University of Louvain, Belgium.

World Disasters Report 2016 Resilience

Saving of lives today and investing for tomorrow, this year's World Disasters Report focuses on resilience within humanitarian action. The report explores the different meanings of resilience, as well as criticisms, and its application both before and after crises.

How resilience is measured and what is the importance of building evidence the report examines the 'business case' for adopting a resilient approach, arguing that investing in resilience yields financial and social benefits, as well as saving lives. It looks at anticipation, and some of the tools and approaches that exist. The report explores the importance of mental health and psychosocial support for those affected by crises, and makes the case that more effort is needed in this often-underestimated area. It reviews the nature of partnership between key actors, and argues that more attention is needed in building more equitable relationships based on trust and transparency. Finally, the report considers the future role of a resilience approach in the face of threats including climate change, conflict and violence.

Salient Features of the World Disasters Report 2016

The Report stresses on the resilience time to act for it, the inner resilience: men health and psychosocial support, stronger together: partnerships that build resilience – resilience in the future: 2025 and beyond – disaster data. More vulnerable people live more exposed to more extreme weather. Tens of millions have been displaced by disasters in recent years. There is no greater challenge for this generation and the next than to prevent and prepare for the brutal force of climate variability and change. It is also a question of justice: the poor who did nothing to produce climate change are first and hardest hit. We, the rich and robust, who caused climate change are last and least hit. The World Disasters Report describes the monumental challenges at hand, but also what is needed to meet them. Jan Egeland, Secretary-General, Norwegian Refugee Council. A Natural disaster is a natural phenomenon, their occurrence is a part of environmental cycle. Common natural disasters are flood, earthquake, landslide, volcanic eruptions, forest fires etc. We can't stop them rather we can mitigate the effects by planning and practice. Occurrence of various natural events is a part of nature, but due to anthropogenic interferences the frequency of disaster has been increased many fold in the present century. The global warming, climate change and land use changes are key factors for the increasing frequency of these disasters as a result of which losses are increased many times since earlier. (World Disaster report 2016).

World scenario

According to secretary general's, report in 43rd UN general assembly, the main culprits identified for causing deaths among the natural hazards were Earthquakes 50.9%, floods 29.7% and cyclonic storms, volcanic eruptions 1.9%, tsunamis 0.5% and landslides 0.1%. Further the report emphasized that major destructions was caused in the developing regions of the world. Asia, south and west pacific 85.5%. Europe, America and Africa have a small share of 14.5% of the total loss.

Economic losses in these regions are multiplying at the rate of 400% per decade which is an alarming situation. In 2016 total losses in disaster estimated at USD 210 billion. It has also been observed that vulnerability of the poor is much more as they are living and working in vulnerable environment therefore, they suffer the maximum damage.

Natural disasters are natural phenomenon but in last two decades human activities are responsible for increasing frequency of these events up to great extent. Humans are changing physical and chemical properties of the atmosphere due to which imbalances are created which leads to global warming, climate change etc. Global warming is raising the sea level and in future it will further increase as global temperature is rising which may bring more frequent El-Nino and La Nino effects. Gradual changes in the climate are manifesting extreme weather conditions causing more hurricanes, cyclones, droughts, floods and storms.

Deforestation is making environment more fragile and exposed to disasters. Urbanization is accelerating without proper norms and planning proving dangerous for mankind during earthquakes, landslides and other hazards.

During the last decade, globally there were 2557 natural disasters which took life of 6, 65, 598 people. Famine killed 2, 80, 000. Out of total casualities by natural disasters, 83% were Asians. Total damages account for 78.7 billion per year worldwide.

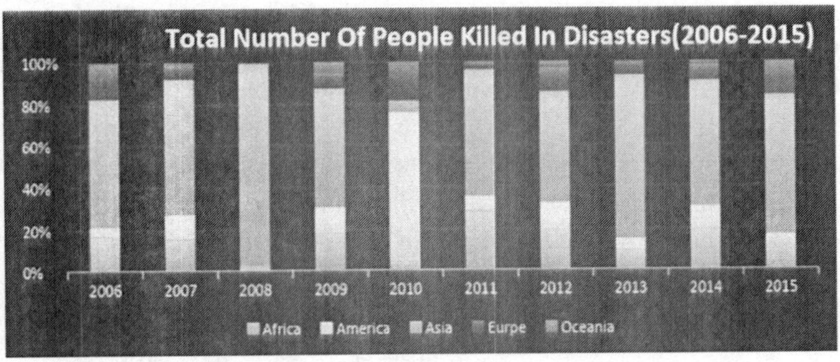

Source of Data: EM-DAT, CRED, University of Louvain, Belgium

Fig. 2.

If we do not change our activities then climate change will manifest serious challenges. As climate is rapidly changing from the past it will continue to bring subtle shifts in average temperatures, rainfall and sea levels. Climate change is aggravating weather phenomena like rainfall snow melting etc. which are converting into the disastrous situations, as a result the frequency of disasters has increased manifold in the present decade. Disasters like the frequency of floods is increasing day by day and it will affect the coastal areas more severely. Where about three billion people that is about half of the total population of the world is living. Thirteen of the world's largest urban centers are located on the coasts. Sea level is rising and inhabitants of coastal zones are becoming vulnerable to flooding. Sea level is expected to rise up to 44 cm by 2080.The biggest threats are to river deltas and small low lying islands. Large vulnerable population live on river deltas in India, Pakistan, Mozambique and Vietnam. Of the total losses floods account for US$ 30 billion in 2012.

Important natural disasters are:

1. Wind related hazards such as cyclones, Floods, Drought, Snow Avalanches, Heat waves, forest fires, etc.

2. Geo physical hazards such as Earthquakes, Landslides, Tsunami, volcanic eruptions etc.

Cyclones

The tropical low pressure storms are known by various names in the world they are called typhoons in the China Sea, in the Indian Ocean they are called cyclones. A cyclone is an intense storm that originates over warm tropical oceans and is characterized by low atmospheric pressure, high winds, and heavy rain, these storms draw energy from the sea surface and maintain its strength as long as it remains over warm water. A tropical cyclone can generate winds that exceed 119 km (74 miles) per hour. In extreme cases winds may exceed 240 km/hour.

They are often accompanying the strong winds and torrential rains and devastating storm surges, an elevation of the sea surface that can reach up to 6 m (20 feet) above normal levels. Such a combination of high winds and water makes cyclone a dangerous phenomenon for coastal areas in tropical and subtropical areas of the world. Every year during the summer months July–September in the Northern Hemisphere and January–March in the Southern Hemisphere, cyclones strike regions along the Eastern Coast of North America, northwestern Australia, and eastern and western coast of India and the coast of Bangladesh.

Tropical cyclones are compact and circular storms, generally 320 km (200 miles) in diameter, winds swirl around a central region of low pressure, and winds are driven by this low-pressure core and because of the rotation of the earth, which deflects the path of the wind through a phenomenon known as the Coriolis force. As a result, tropical cyclones rotate in a counter-clockwise direction in the Northern Hemisphere and in a clockwise direction in the Southern Hemisphere.

The structure of a cyclone

Eye is a characteristic feature of tropical cyclones, a central region of clear skies with warm temperatures and low pressure. Atmospheric pressure at the surface of earth is generally goes to about 1, 000mb at the Centre of a tropical cyclone, it is typically around 960 mb, and even low, in a very intense super typhoon it may be as low as 880 mb. In addition to low pressure at the Centre, the pressure varies across the storm, with most of the variation occurring near the Centre. This variation results in a large pressure gradient force, which is responsible for the strong winds present in the eye.

Winds within the central region, are light. In addition to this there is a weak sinking or subsidence, as air is pulled into the eye because of low pressure. As the air subsides, it compresses slightly and warms up, so that temperatures at the Centre of a tropical cyclone are around 5 °C (10 °F) higher than in other regions of the storm. The eye of the cyclone is mostly free of clouds. The most dangerous and destructive part of a tropical cyclone is the eye of the cyclone here winds are strongest, rainfall is heaviest, and deep convective clouds rise from close to Earth's surface to a height of 15, 000 m or 49, 0 00 feet. As noted above, the high winds are driven by rapid changes in atmospheric pressure near the eye, which creates a large pressure gradient force.

The upward movement of air in the eye also causes the eye to be wider aloft than at the surface, the wind speed decreases with height, the air must move farther from the centre of the storm as it rises and reaches the stable tropopause the upper boundary is at 16 km [10 miles] above the surface, the air flows outward. The Coriolis force deflects this outward flow, creating a broad anticyclone circulation aloft. Therefore, horizontal circulation in the upper levels of a tropical cyclone is opposite to that at near the surface.

Rainfall

In the central part of the cyclone vertical air movement takes place, sometimes rain bands are arranged around the centre. The rotating cloud bands are often associated with wobbling of the storm track, it can cause large differences between the forecast and actual landfall.

As a tropical cyclone strikes the coast, surface friction increases which in turn increases the convergence of airflow into the eye and the vertical motion of air starts there. The increased convergence and rising of moisture-laden air is responsible for the torrential rains associated with tropical cyclones, which sometimes exceeds 200 mm (9 inches) in a day.

Life of a Cyclone

A circulation system goes through a sequence of stages as it intensifies into a mature tropical cyclone. The storm begins as a tropical disturbance, which typically occurs when loosely organized cumulonimbus clouds in an easterly wave begin to show signs of a weak circulation.

The conditions favorable for this process to take place are:

1. The surface temperature of the ocean water must be 26.5 °C or high, this layer should be 50 m deep.
2. Near the surface is the warm layer where an atmospheric circulation must be located.
3. The normal lapse rate of temperature with height is required to support the formation of deep convective clouds.
4. The atmosphere must be relatively humid at a height of about 5, 000 m (16, 000 feet) above the surface.
5. The development of cyclone should take place away from the Equator, the energy or heat for a tropical cyclone is provided by transfer of water vapour heat.

Intensification of a cyclone

The dynamics of a tropical cyclone depends on the fact that the external part of a storm being cooler than its core, so it is necessary that the temperature of the atmosphere drop rapidly with height. The warm and saturated air rising in the centre of the circulation tends to keep rising as long as the surrounding air becomes cooler and heavier.

Among all cyclones hurricanes are the most destructive, they claim a higher share of deaths and property losses. According to one estimate about 80 tropical cyclones are formed all over the world. India is frequently affected by tropical cyclones because of its vast coastline, 13 states and 84 districts are vulnerable to this hazard, tropical cyclones cause heavy loss of life and property in India, Bangladesh, Sri Lanka, China etc. Cyclones generally occur between April and May and October – December.

Some important cyclones in the recent past are cyclone Nargis, Hudhud, Phaillean, cider etc. Cyclone philean was a very severe cyclone which made its landfall in India on October 14, 2013. Indian Meteorological Department warned in advance that a cyclone is going to strike on Indian coast, Indian government evacuated about 4.4 lakh people from the vulnerable areas, largest in the Indian history, due to which human life loss was reduced to minimum and only about 50 people died and millions were saved but huge property loss was incured.

On May 3, 2008 a severe cyclone Cider struck Myanmar and 138, 368 people died. It is the fifth deadliest cyclone disaster on record, causing property damages worth $12.9 billion and it affected Bangladesh, Myanmar, India, Sri Lanka, Thailand, Laos, and China.

Cyclone Hud Hud a severe cyclone originated in the Andaman Sea, caused extensive damage to the city of Vishakhapatnam in India. Property worth $ 43.4 million was lost and 154 people died, it affected Andaman Nicobar, Andhra Pradesh, Madhya Pradesh.

Hurricane Gustav-made landfall in Cuba on 30 August 2008, causing damage of $210 million and 154 people died in Jamaica.

Mitigation

Cyclones cannot be controlled, but through proper mitigation strategies and policies such as installation of early warning systems, such systems are fixed along the coastline greatly assist forecasting techniques, therefore it helps in early evacuation of people in the storm surge areas.

The mitigation measures which can help in effective disaster management for cyclones are as follows:

Helipads should be constructed at regular intervals through highways and roads. Sites for air landing should also be available for vulnerable locations such as schools, playgrounds etc.

Support should be provided to amateur radio volunteers at times so that they can effectively handle the emergency conditions caused by landfall of a cyclone. Regular Maintenance of embankments should be done with good workmanship to mitigate risk of flooding during cyclones. There should be provision of Incentives to reward states for good disaster preparedness. Long term mitigation measures should also be in place in district plans in coastal towns and cities, building rules need to be revised and structural safety is a must, all the structures constructed should be cyclone resistant.

Infrastructure in vulnerable areas such as roads, public buildings, bridges, communication systems and key infrastructure such as airports and hospitals must be resilient to cyclones.

Ageing test and assessment of retrofitting requirements should be mandatory for major city buildings which are more than 15 to 20 years old, Coastal Regulation Zone norms should be strictly followed, Coastal Zone Regulatory Authority should be created with state counterpart organizations covering the entire coastline of the countries for management of marine resources and reserve forests. Coastal Planning as well as balancing of the coastal vegetation and green cover should be taken up.

Suitable guidelines should be issued by the states for construction of hoardings. Evacuation routes in disaster should be delineated in the management plans prior to onset of monsoons. Telecommunication systems need to be developed wherein coastal mobile towers must be able to bear winds at speeds of 250 km/hr.

Alternative communication channels should be set up including VF/HUF sets, satellite phones, radio, internet and loud speakers for communication during the crisis.

Underground power cable and utility lines should also be planned in a manner that they are functional and economically viable last mile connectivity must be ensured.

Important Facts

Early warning systems such as IMD, CWC, and INCOIS are the key agencies for disaster management when a cyclone strike. Advisory service providers during cyclones include:

1. Agro met Advisory Service (IMD)

2. Potential Fishing Zone Advisory Services (INCOIS)
3. Agriculture advisory service (Department of Agriculture)

Other stakeholders include the following:

Road, rail and civil aviation authorities. Disaster Management Agencies and Defense, police, power, telecom, health, water and sanitation. Educational departments. Print and Electronic Media, NGOs and organizations of public and private agencies across coastlines.

Development of communication system

Communication plays key role in cyclone disaster mitigation, radios, transistors, television etc. can be used as an important tool for disaster mitigation.

Shelter belts can make by plantation of trees along the coast can act as wind breakers. They can protect soil erosion and prevent damage of crops during disaster.

Building up of Cyclone shelters

Cyclones shelters are built in the cyclone prone areas to minimize the loss of human life. In the normal days these shelters are used as public buildings.In the cyclone prone areas, especially designed concrete houses should be built which can withstand high winds and tidal waves. Public awareness program should be organize to inform the population of vulnerable areas have to response cyclone warnings to reduce the losses.Any settlement or industrial buildings should not be permitted in the coastal belt of 5km from the sea, which is most vulnerable area as far as cyclones are concerned, settlements and all the important establishments should be built beyond 10 Km from the sea.

Cyclones Warnings

The Geo-Stationary satellites are provide continuous data on day to day basis for the weather elements, SPOT, GOES, INSAT, IRS, are among those satellites. Cyclones can be predicted with high accuracy through cyclone tracking since its origin, every point is explained since the beginning where a cyclone depression would formed, from where it pass and where it will strike. This help in taking preventive measures like evacuation if known earlier. There are many instances such as cyclone phillean, early warnings were issued 5 days before its arrival, Indian government took initiative and about 4.4 lakh people were evacuated safely. As a result life loss was minimum. A number of instance are there in the world where early warnings play vital role in cyclone mitigation.

Table 3: Cyclones in the world

Country	*Year*	*Place of origin*	*Death*
Bangladesh	1970	Bay of Bengal	3-5 Lakh
China	1975	West Pacific	171, 000
Bangladesh	1991	Bay of Bengal	138, 366
Myanmar	2008	"	138, 336
Bangladesh	1985	"	15, 000
India	1977	"	14, 204
Bangladesh	1965	"	12, 047
India	1935	"	60, 000

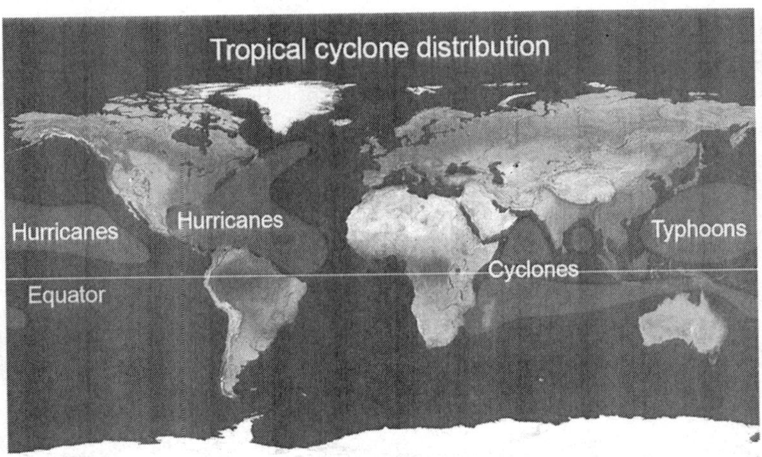

Source: Met o office. Gov. UK, Tropical cyclone Map.

Earthquake

An Earthquake is caused by sudden release of energy from the earth's crust. Earthquakes are mainly associated with release of energy from the interior of the earth which cause damage of structures on the surface of the earth, therefore major destruction cause by collapse of buildings if buildings are not there losses are minimum. Earthquake also occur within the tectonic plates such as in china, Russia, South west United States.

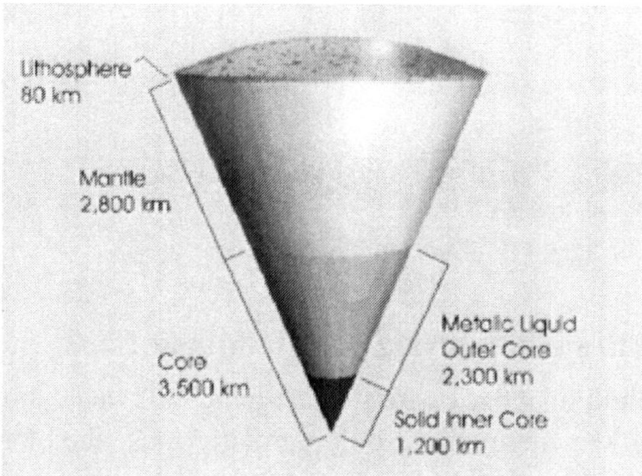

Fig. 3: Interior of the earth credit USGS.

What causes earthquakes?

The interior of the earth has four major layers:

1. The inner core
2. outer core
3. mantle and
4. Crust (figure 4).

The crust and the top layer of the mantle make up a thin layer on the surface of our earth. But this layer is not one piece it is made up of many pieces like a puzzle covering the surface of the earth. These blocks keep moving slowly, sliding past one another and riding into each other these pieces are called tectonic plates, margins of the plates are called the *plate boundaries*. The plate boundaries are made up of many faults, and most of the earthquakes around the world occur on these faults edges of the plates. The plate margins are rough, they get stuck while the rest of the plate keeps moving.

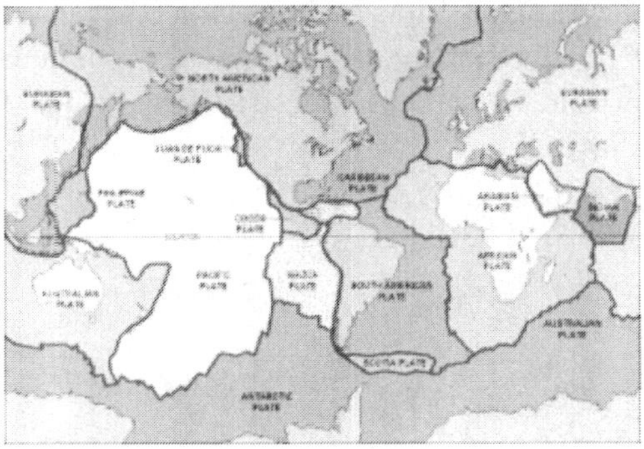

Source: USGS.

Fig. 4: Tectonic Plates.

Why earth shakes during an earthquake?

The fault boundaries are stuck together and the rest of the block is moving, the energy that would normally cause the blocks to slide past one another is being stored up. As soon as the force of the moving blocks finally overcomes the friction of the jagged edges of the fault

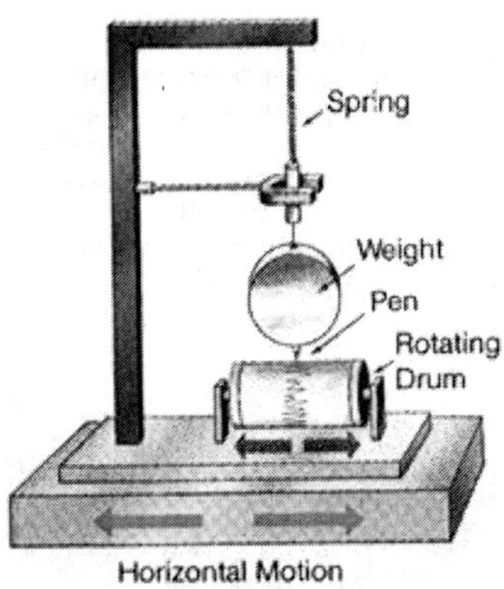

Source: USGS.

Fig. 5: Recording instrument of earthquakes.

and it unsticks, all the stored up energy is released and radiates outward from the fault in all directions in the form of seismic waves like ripples on the water surface. The seismic waves shake the earth as they move through it, and when the waves reach the earth's surface, they shake the ground and everything on it.

Recording of the earthquakes

Earthquakes are recorded by instruments called seismographs. The recording they make is called a *seismogram.* (figure3) The seismograph has a base that sets firmly in the ground, and a heavy weight that hangs free. When an earthquake occur the ground shakes, the base of the seismograph also shakes, but the hanging weight does not. The spring that it is hanging from absorbs all the movement. The difference in position between the shaking part of the seismograph and the motionless part is what is recorded.

Measuring of the size of the earthquakes

The size of an earthquake depends on the size of the fault and the amount of slip on the fault, but that cannot be measured simply with a measuring tape since faults are many kilometers deep beneath the earth's surface. They use the *seismogram* recordings made on the *seismographs* at the surface of the earth to determine how large the earthquake was. A short wiggly line that doesn't wiggle very much means a small earthquake, and a long wiggly line that wiggles very much means a large earthquake the length of the wiggle depends on the size of the fault, and the size of the wiggle depends on the slip of rocks.

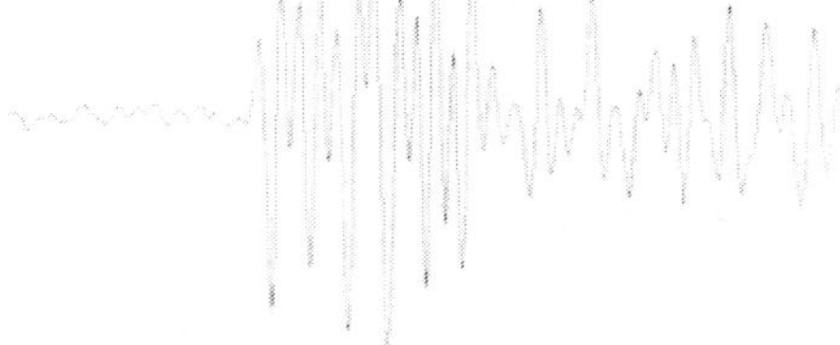

Source: USGS.

Fig. 6: Siesmograph.

The size of the earthquake is known as its magnitude. Each earthquake has a magnitude scientists also talk about the intensity of shaking from an earthquake, and it varies depending on where we are during the earthquake.

When there is an earthquake three types of waves are released P, S and L waves.

P waves are faster than S waves, and on this basis we can tell where an earthquake was. The P waves travel faster and shake the ground where you are first. Then the S waves follow and shake the ground also. If you are close to the epicenter, the P and S waves will come one right after the other, but if we are far away, there will be more time between the two. By looking at the amount of time between the P and S wave on a seismogram recorded on a seismograph, scientists can detect how far away the epicentre was from that location.

However, they can't tell in what direction from the seismograph the earthquake was, only how far away it was and if they draw a circle on a map around the station where the radius of the circle is the determined distance to the earthquake, they know the earthquake lies somewhere on the circle.

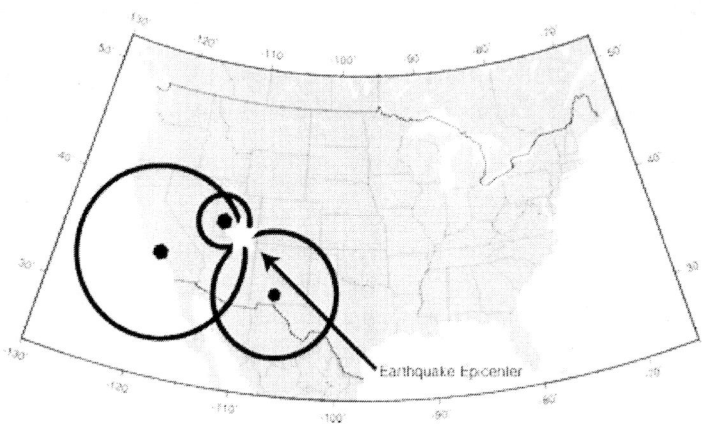

Source: USGS.

Fig. 7: Triangulation Method for earthquake detection.

The fig. shows triangulation method to determine the location of earthquake.

Scientists then use a method called *triangulation* to determine the exact location of earthquake (figure 7). It is called triangulation because a triangle has three sides, and it takes three seismographs to locate an earthquake. If you draw a circle on a map around three different seismographs where the *radius* of each is the distance from that station to the earthquake, the intersection of those three circles is the epicenter.

Can earthquakes are predictable

Scientists are still not able to predict an earthquake. Scientists have tried hard to find out many different methods of predicting earthquakes, but till date none have been successful. All the faults can become active at any time, earthquake can occur in the future, but we have no mean of predicting that when it will happen.

Is animal behaviours and some weather clues can warn us before an earthquake occur

About ten to fifteen hours before the quake, most of the animals, birds, reptiles, insects etc become extremely uneasy and make unusual noise and shout (Bapat.A). This has been seen prior to all earthquakes. They are in fearful conditions and their movements are in haphazard directions. A number of people are also disturbed with some unusual disturbing feeling and in some cases the blood pressure rises without any reason. Diseases such as heart trouble, blood pressure, respiratory disorders, and nervous diseases are aggravated.

About ten to fifteen hours before the quake the television reception suffers from audio, visual and spectral disturbances. Any amount of repairing of the fault does not help in getting the correct picture on the TV screen. Such effects are more pronounced in antenna reception and somewhat less in cable connection. The reception on radio and telephone is disturbed and there are continuously disturbing noise.

About two to three hours before the quake most of the mobiles and telephones start mal-functioning or non-functioning. The Bhuj

Earthquake occurred at 0846 hours and the above effect was seen from about 0600 hours.

The observations from Peru earthquake of 15 August 2007 and China earthquake of 12 May 2008 show that about a sudden glow in the sky at a short height. Such glows are seen about thirty minutes before the quake. Seismologists describe the glow as 'earthquake glow'. It is seen before the quake whether it is daytime or nighttime. Japan and China have done remarkable research on this subject.

Source: Epoch times.

Fig. 8: Earthquake glow.

This photo was taken at 13:34:20 on May 12, 2008 at the Gaoxin 4th Road, East Development Zone, Baoji City, Shaanxi Province.it shows the glow appear on the sky above the region before a big earthquake.

World Distribution of Earthquake

There are four belts of active seismicity in the world they are:

(a) Young Folded Mountain Regions.
(b) Regions of faulting and fracturing.
(c) Active volcanic regions, and
(d) Along the margins of different plates

(1) Circum –Pacific belt -about 70 % of the earthquakes of the world occur along the "Pacific Ring of Fire".

(2) Mid Continental belt- about 20% of the world's earthquakes-occur here it includes Mediterranean-Himalayan belt.

(3) Mid Atlantic Ridge- According to one estimate around 10 % of the total earthquakes including the earthquakes of new Madrid, Charleston, Boston and Konya etc. are originate in this region.

The dangers of earthquakes are profound and the damage they do is catastrophic. Their effects are of two types- primary and secondary. Primary effects causes damage directly and include ground motion and faulting; secondary effects causes' damage indirectly as a result of process set in motion by the earthquake.

Causes of earthquake

Causes of earthquakes are various but main cause is the release of energy for the interior of the earth. Presently we know where and how the earthquakes are being originated to a great degree of precision. We still fail to estimate accurately is their future behaviour. Prediction and warnings of earthquakes is still a challenge in spite of detailed temporal and locational records, a dense global network of seismic stations and hazard maps based on real time data and support of sophisticated monitoring system, many case studies of earthquakes occurred so far revealed that reasons behind them vary from one place to another.

Plate tectonics and seismic activity

Large continents and ocean blocks, long continent-continent and oceanic-oceanic convergence earthquakes are comparatively not so frequent or large. Where two continental margins collide along their margins, the focus can be up to 200 km deep. The best example is the belt running through the Himalayas, Iran, Iraq, Black Sea, Caspian Sea and the Mediterranean Sea.

1. Divergent plate margins and seismicity

When two plates move away from each other cause divergence

and tension, these boundaries show low tremors when measured on Richter scale the focal depth lies within 20 km, because crust here shows thinning. The East African rift valley, Basin and Range province of Western North America and along all mid-oceanic ridges like Mid Atlantic Ridge and East Pacific Rise are the characteristic examples.

2. Convergent plate margins and seismicity

When plates slide and past each other, they show only horizontal movements. Along such margins, if blocks move constantly they do not store strain. These blocks are unlocked blocks and their displacement is called creep, which is a seismic converse, blocks are structurally joined, these blocks are called locked blocks. **San Andreas Fault** shows both types of blocks these blocks have elastic properties and they store the force applied on them by deforming the rocks till they cross their last limits of elasticity. Rocks slip like a rubber band try to regain their original configuration all the energy stored in them is released at the time of slippage then main shocks are occurred at this time, while the region records foreshocks prior to the main shocks and aftershocks later the aftershocks spread over a larger area as the rocks comeback to their original shape. The most prominent examples of trans-current plate margin earthquake zones are the San Andreas Fault and the South Island of New Zealand. These earthquakes have shallow focus, less than 100 km deep. The tremors are of moderate magnitude on the Richter's scale.

Conservative margins

Convergent margins are also known as transform fault boundaries. These are mainly found on the ocean floor, where they offset mid ocean ridges and enable to ocean to spread at different rates and these faults were recognized as the connection between the ocean ridges (divergent margins) and ocean trenches (convergent margins) by scientist John Tuzo Wilson.

At conservative margins plates slide past each other, so that the relative movement is horizontal. Lithosphere is neither created nor sub ducted, and whilst conservative plate margins do not result in volcanic activity and they are the sites of extensive shallow focus earthquakes, occasionally of considerable magnitude.

Some of the deadliest earthquakes of the world are:

- July28, 1976, Tangshan earthquake China, 700, 000 people killed, and the magnitude of earthquake was 7.8.
- Indian Ocean earthquake 2004, an earthquake with a magnitude of 9.3 was occur in Phuket Indonesia create devastating tsunami about 230, 210 people were died as a result great havoc was created in all the Indian Ocean nations.
- Haiti earthquake January 12, 2010, an earthquake with magnitude 7.0 was occur, around100, 000-316, 000 people were killed.
- Kashmir earthquake October 8, 2005, Muzaffarabad Pakistan an earthquake rock the region killed 86000-87351 were killed, 138, 000 were seriously injured and 3.5 million were displaced.
- Japan earthquake occur in Canto Plain of Honshu Island devastated Tokyo, Yokohama, Chiba etc. The magnitude of the earthquake was 7.9.

Earthquakes are not predictable till date this is the reason they cause heavy damage of life and property. There is no early warning in case of earthquakes it can happened any time. Only Seismic hazard zonation is possible on the basis of analysis of past earthquakes and presence of fault lines is a well-known method for locating dams, power plants and other projects in seismic zones. Land sat-TM and SPOT images and Radar interior programs have been used to detect the active faults. Side-Sonar system can be used to detect active faults on the seafloor: Earthquake prediction is experimental stage. Only few earthquake in the world are satisfactorily predicted, for example on earthquake was predicted 1-2 days ahead in 1975 (Vogel, 1980).

Information regarding the earthquakes is collected from seismographs. Now space geodetic techniques and high resolution aerial and satellites data have been used for earthquake prediction.

Surface deformations can be easily measured with an accuracy of a centimeter over 100km, prior to earthquake.

Earthquake risk assessment done on the basics of identification of earthquake zones through collection of geological, physical and geomorphological data and mapping of known seismic activities in the region.

Satellite imagery could be used in the delineation of geological structures and to analyse seismological conditions in earthquake risk zones. Mapping of lineaments reveal active movement and tectonic activity along the faults. A close relationship activity is* observed in Latur area of Maharashtra. GPS is extremely useful in measuring the complex deform patterns.

Across plate boundaries where strains are build up. Slip along faults have been measured with GPS to an accuracy to sub-centimeters.

Table 4: Deadliest Earthquakes of the World

Place	*Year*	*Magnitude*	*Death Toll*
China	1976	7.8	700, 000+
Indian Ocean Sumatra	2004	9.1-9.3	230, 210+
Haiti	2010	7.0	100, 000+
Kashmir India	2005	9.1	87, 351
Bhuj Earthquake	2001	7.7	13, 805

Source: world disaster statistics.

Mitigation

Earthquakes are not predictable so early warnings cannot be issued only we try to reduce the economic and social losses for that a comprehensive risk assessment of seismic hazard, through increasing awareness, land use.

Planning according to the risk is seismic zones, formulation of building codes to prevent buildings during earthquakes.

GIS can be used to correlate the relationship among after shock epicenters to know about the interior of the earth. In a study Gupta et.al. (1954), GIS was used to integrate geological, topographical and seismic information to generate an isoseismic map of the Uttrakhand.

Earthquake is a geo-physical hazard, causes complex ground motion. This motion is both vertical and horizontal as a result the structures vibrate. The failure of the structural elements make the structural frame of buildings. A vicious cycle of load transfer mechanism is a pre-requisite.

Earthquakes are unpredictable, to reduce its impacts only solution is its prevention.

All over the world about 75% of fatalities attributed to earthquake are caused by collapse of buildings but in future it will decrease due to new designs and building codes. Developing countries have large population so they are more vulnerable to earthquake as compare to developed countries. The developed countries spent enough resources on mitigation of earthquake resistant technology and norms of building quotes engineered buildings are designed on the basis of seismic codes and gravity load to resist earthquake.

Buildings built without engineered design are week, most susceptible to danger of earthquakes, they are vulnerable to seismic hazard.

In case of low intensity earthquakes structures remain unaffected, event nonstructural buildings should suffer any damage.

In the context of a medium intensity a perfect elastic design of structure, some minor damage is permitted, the structures have high degree safety against collapse.

In the event of a most severe earthquake structures are built against collapse and save human life lines.

Earthquake protected shapes of buildings

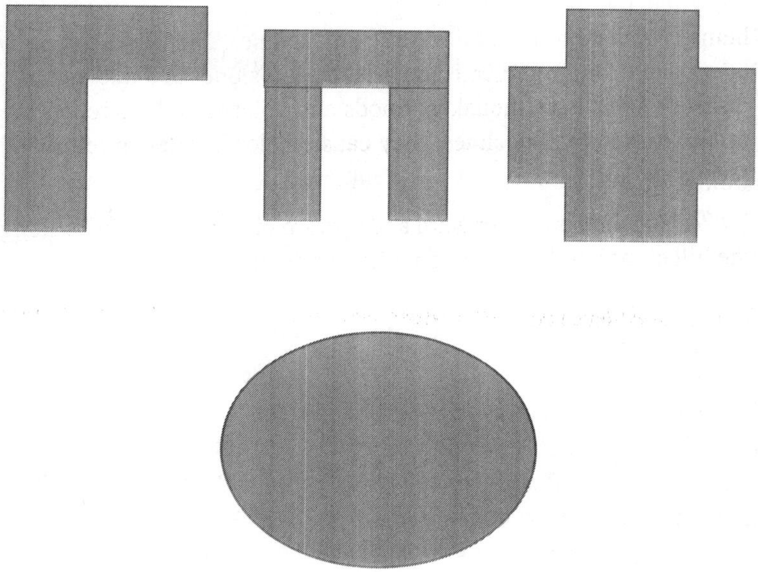

Fig. 9: Earthquake protected shape of the buildings.

How we can protect one buildings by choosing correct designs?

Generally the weak sections of the buildings are joint of walls, portion of the wall between two openings, type of roof etc.

House can be made earthquake resistant in the following way.

Vertical steel must be use near the openings and at the corners and junctions of the walls to increase ductile behavior so that buildings can resist the tension deforms bend instead of collapse.

Stone can be used in walls at regular intervals to make it more resistant. Heavy roof made of store is more prone to damage.

To develop a safety culture it is necessary that we implement the available technology and develop an earthquake resistant infra-structure and a habit of use of safety norms of construction among artisans, masons builders, architects and engineers.

Landslides

Landslide is a major geological hazard, poses serious threat to human population and for infrastructure facilities. Landslides generally occur due to anthropogenic activities on hill slopes and due to major disasters such as earthquakes, floods and volcanoes, land routes are often blocked by landslides, they cause major hurdles in relief and reconstruction work.

A landslide is a downward and upward movement of rocks along the hill slopes under the action of gravitational force.

Causes of landslides

Factors responsible for landslides are both anthropogenic and natural.

I. Natural factors are, the inherent characteristics of rocks and hill slopes like geology, gradient of slope, relief, hydrology of the region, land use and land cover. Landslides can occur as a result of some disastrous event such as heavy rainfall, cloud bursts, volcano eruption, earthquakes, etc.

II. External factors include atmospheric or environmental factors such as agent of denudation, chemical and mechanical weathering, and deforestation. Frost effect causes freeze and thaw action and create crevasse in the rocks, which weakens the rocks by widening of joints and fractures hence rocks breakdown and slides.

III. Geology: permeable rocks with joints and fractures because of intense folding and faulting become weak and prone to slide. An increase of shear stress and decrease of material strength.

IV. Change in the gradient of the slope: undercutting of slopes by rivers and streams causing steep bank slopes. High relief along with steep slopes generally have high potential for the rock fall.

V. Change in the gradient of the slope: undercutting of slopes by rivers and streams causing steep bank slopes. High relief along with steep slopes generally have high potential for the rock fall.

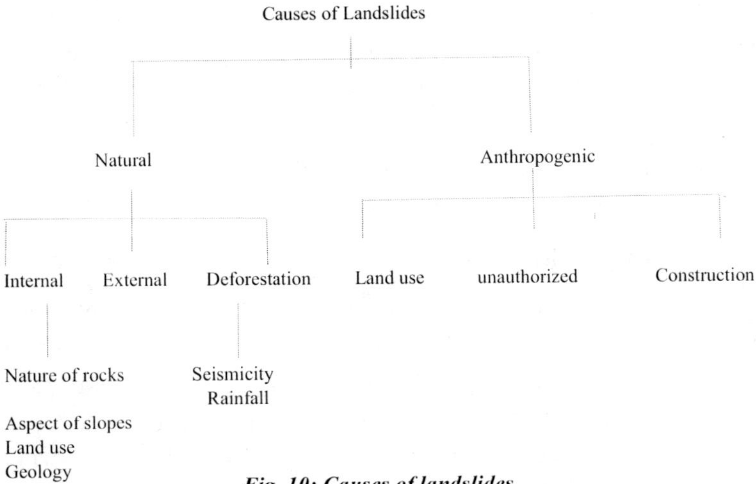

Fig. 10: Causes of landslides.

Classification of Landslides

Varner proposed a very comprehensive classification in 1978, the classification is based on type of movement and type of material involved, into fall slides creep, debris flow and earth flow etc.

Some devastating landslides of the world includes numerous landslides in Sikkim and West Bengal killed thousands of people due to Darjeeling floods in 1968, these landslides occurred due to three day heavy precipitation ranging from 500 to 1000mm.The highway to Darjeeling was blocked at 92 places resulting in disruption of communication system.

Alaknanda episode of July 1970 is another example when due to massive floods in river Alaknanda

In Malpa tragedy whole village was wiped out instantly killed 220 people on the right bank of river Kali in the Kumaun Himalayas.

Amboori landslide in 2006 in Kerala 23 people were killed due to debris flow consisting of boulders.

North Indian floods 2013 when a series of floods and landslides ravaged parts of the mountainous state of Uttrakhand more than 5700 people died rivers are blocked by falling debris.

Gansu Mudslide 2010: This landslide was triggered because of heavy rainfall in China claimed 1, 471 people multi story buildings were collapse.

Guatemala, October 1, 2015 heavy rain resulted in a massive landslide. More than 200 people died and several went missing.

Bangladesh landslide 2017- A series of landslides in three hilly districts of Bangladesh killed more than 152 people heavy rain triggered the landslide.

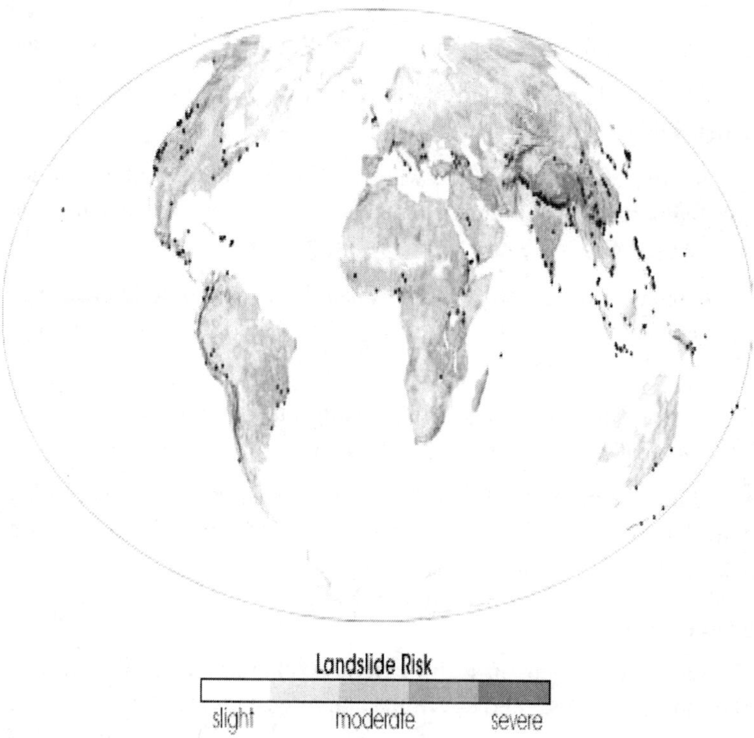

NASA map by Robert Simmon, adapted from Hong, Y., Adler, R., and Huffman, G. (2006). Evaluation of the potential of NASA multi-satellite precipitation analysis in global landslide hazard assessment.

Varner proposed a very comprehensive classification in 1978, the classification is based on type of movement and type of material involved, into fall slides creep, debris flow and earth flow etc.

Some devastating landslides of the world includes numerous landslides in Sikkim and West Bengal killed thousands of people due to Darjeeling floods in 1968, these landslides occurred due to three day heavy precipitation ranging from 500 to 1000mm.The highway to Darjeeling was blocked at 92 places resulting in disruption of communication system.

Alaknanda episode of July 1970 is another example when due to massive floods in river Alaknanda.

In Malpa tragedy whole village was wiped out instantly killed 220 people on the right bank of river Kali in the Kumaun Himalayas.

Amboori landslide in 2006 in Kerala 23 people were killed due to debris flow consisting of boulders.

North Indian floods 2013 when a series of floods and landslides ravaged parts of the mountainous state of Uttrakhand more than 5700 people died rivers are blocked by falling debris.

Gansu Mudslide 2010: This landslide was triggered because of heavy rainfall in China claimed 1,471 people multi story buildings were collapse.

Guatemala, October 1, 2015 heavy rain resulted in a massive landslide. More than 200 people died and several went missing.

Bangladesh landslide 2017- A series of landslides in three hilly districts of Bangladesh killed more than 152 people heavy rain triggered the landslide.

Prevention

Landslides occur in hilly regions so there is a need of careful planning of land use land cover, the activities which make rocks weak and cause slope instability, catchments of river channels building of structures on fragile slope must be avoided, increase afforestation on the hill slopes to reduce soil erosion and weathering of rocks, settlements and other important land use must be avoided in landslide prone areas. In India whole Himalayan region and some regions of south India are prone to landslides.

The most deadliest landslide of the world till date are reported from all over the world are

The multi temporal images from Landsat TM and MSS can be used to ascertain relationship between the land use and the occurrence of landslides in various parts of the world.

Table 5: Annual Losses by Landslide in the World

Country	*Yearly losses Million US $*
Japan	4.700
Italy	2.600
United States	1.800
India	1.350
China	500
Spain	220
Canada	50
Hong Kong	25
New Zealand	12
Norway	6

Table 6: Major Landslides in the world since1960

Country	*Year*	*Deaths*
China	1933	3000+
Tajikistan	1949	4000
Peru	1962	4500
Indonesia	1919	5000+
Columbia	1985	23000
Venezuela	1999	30, 000
China	1920	100, 000
Kedarnath(India)	2013	5700
PERU	**1970**	**22000**

Source: World Disaster statistic.

Monitoring

Monitoring of unstable slopes using instruments are employed in major engineering project sites, river valley projects, colonies important roads and rail routes. Data obtained has a greater significance as it provides data related to slope deformation pattern, but it involves very high costs. Two type of monitoring is done:

1. Superficial monitoring using precision levelling of survey pillars located within the landslide prone areas.
2. Subsurface deformation monitoring of slopes.

Community based mitigation

Most damaging landslides occur due to human intervention such as construction activities to mitigate hazard education and awareness among the community can play an important can play an important role in reducing the accidents. This would also reduce the burden of shifting of settlements.

Snow Avalanches

Snow Avalanches are a common phenomenon in the snow bound mountain regions of the world it is a slide of snow down a mountain side under the effect of gravitational pull.

This is a down slope movement of a large detached mass of snow and ice, along with debris such as rocks and flora, small avalanches occur in large number. Large avalanches that may encompass with millions of tons of snow accumulated in frequently and is the cause of damage.

An avalanche has three main parts.

1. The striating zone- is the most vulnerable area of slope, from here unstable snow detached and begin to slide.
2. The avalanche track is the path of the snow through which goes down piles of snow and debris, at the bottom of vertical slope indicates that an avalanche have run.

3. The Run out zone- is that place where snow and debris are stopped, here deposition takes place and in this region snow and debris piles the highest.

Avalanche site

An avalanche can go for many kms and resulting massive loss of flora and anything else which come its way.

In high mountainous areas these avalanches are very common. Higher Himalayas, Hindu Kush, Karakoram experience may large avalanches every year. Snow avalanches are commonly occur in the mountains of all continents and climatic zones, the lower limit of avalanche distribution starts from sea level in polar and temperate zones to 6000 m in the tropics. Four altitudinal belts of avalanching are exit in the mountainous areas:

permanent snow

stable seasonal snow cover,

unstable snow cover and rare snowfalls

Most avalanche risk prone areas of the world belong to the stable seasonal snow-cover. Factors that affect an Avalanche are: weather conditions, temperature, slope gradient, aspect of the slope, direction of winds, vegetation and amount of snow.

Types of snow Avalanches

1. Loose snow avalanches
2. Slab Avalanche

Prevention: change in the land use land cover pattern, and afforestation on the hill slopes.

Fig. 11: Snow avalanche in the valley- credit: GNS Kashmi.

The inhabitants of mountainous areas avoid to visit the areas where warnings are released.

Mitigation

Avoid to visit those areas where avalanches are common, in avalanche prone areas track should be identified and precautions can be taken to minimize the losses such as construction of available barriers in avalanche areas are very effective, snow net can be used as barriers they can be constructed with concrete, steel or wood. Avalanche dams can also be made to stop or deflect avalanches. These barriers are made up of rocks or concrete. They are put above the structures one want to save, along transport corridors, large shelters known as snow sheds can be built in the slide path of an avalanche in this way we protect traffic from avalanche hazard.

Early Warnings

Satellites provide data where an avalanche is going to occur, interferometry rocks, high resolution sensors can monitor unstable pockets over a long term basis, past data is analyzed and upcoming ruptures are recognized. These systems can recognize events several days in advance.

Alarm system

Remote sensing and GIS techniques make possible to monitor and forecast avalanches, alarm systems are developed to warn people if there is an avalanche going to occur, such alarm systems are installed in endangered areas, e.g. such an alarm is fitted on the only access road of Zermatt in Switzerland. The system automatically closes the road by activating barriers and lights within seconds.

Rescue and Recovery

Terrain management: involves reducing the exposure of an individual to the risks of travelling avoid under cutting slopes. Awareness: people of vulnerable areas should make aware about the threat and give them related information of does and don'ts to create safety culture,

Leadership: Well defined decision making protocols to observe risk factors must be there.

Equipment: Various search and rescue equipment like avalanche cords, Beacons, probes, shovels and sophisticated equipment like life guards can be used for search and rescue.

Table 7: Deadliest Avalanches in the world.

Country	*Year*	*Deaths*
Russia	2002	125
Pakistan	2012	138
Afghanistan	2010	172
India	1979	254
Peru	1962	4000
Italy	1916	10000
Peru	**1970**	**20000**

Location of glaciers that range in size from ~1 to 5 km

NESDIS / National Snow and Ice Data Center

Tsunami Disaster

Tsunami is one of most destructive natural disasters known to man. Tsunami means harbor waves, as it is observed that these waves increase their height up to several meters when they reached the coast where as in the middle of the sea they proved most devastating. Tsunamis are giant and powerful waves most often caused by earthquakes beneath the ocean floor. Their incredible power can destroy entire communities, then drag the debris out to sea. Tsunamis originate because of underwater earthquake and volcanic eruptions the vibrations travel through the water thousands of kilometers, Thucydides in 426B.C inquired in his book History of the Peloponnesian War about the causes of generation of tsunami waves and was the first to argue that ocean earthquakes must be the cause of Tsunami, which can be generated when the sea floor abruptly deforms and vertically displaces the overlying water. Tectonic earthquakes are a different kind of earthquake that are associated with the Earth's crustal deformation when these earthquakes occur under the sea, the water above the epicentre area is displaced from its equilibrium position. Moreover, a tsunami can be generated when faults associated with convergent (destructive) plate boundaries move abruptly, resulting in water displacement, due to the vertical component of movement involved, movement on normal faults can also cause displacement of the rocks of the seabed, but only the largest of such events in the outer

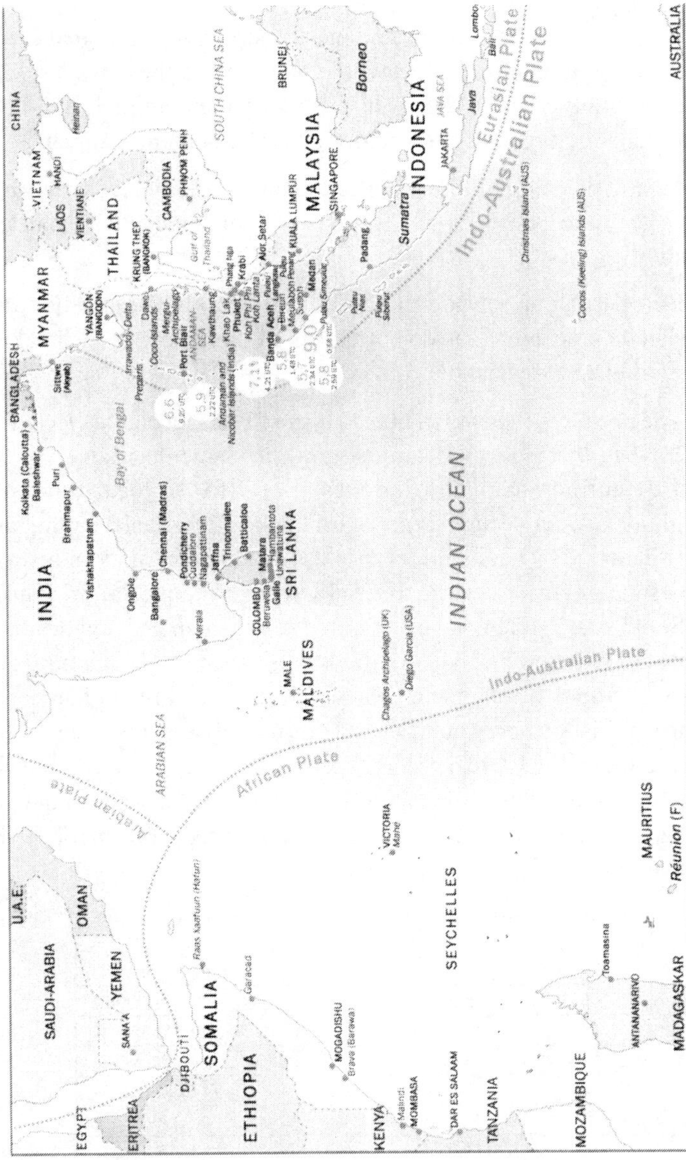

Source: Wikipedia. Countries affected by Tsunami 2004.

trench cause enough displacement to give rise to a significant tsunami, such as the 1977Sumba and Sanriku 1933 events.

This is create the massive and most destructive waves that create huge destructions in the areas where they invade these waves wipe out everything come in their way including men, animal, buildings and other structures. Most tsunamis–about 80 percent–happen.

Within the Ring of Fire, Pacific Ocean a geologically active area where tectonic activities are responsible for huge volcanism and seismicity.

Some tsunamis invade the coast as a quickly surging tide that inundates coastal areas and do not appear on shore as massive breaking waves but instead resemble.

The deadliest Tsunami of human history is that occur on December 26, 2004 in the Indian Ocean just north to Sameulu island, off the coast of northern Sumatra Indonesia, at 7:58:33 local time at a magnitude of 9.0 (second largest earthquake after Alaska earthquake 1964) generated a tsunami which is the deadliest in modern history. This Tsunami create destruction and havoc in Indonesia, Sri Lanka, and South Island, Thailand etc. Caused serious damage and death up to coast of Africa. The height of the waves was up to 15m high. Total twelve countries were effected. At that time none of the Indian ocean country have Tsunami watch centres due to, only two instances of the past are there 1881 and 1941 mild effect was experienced, and for this reason none of the country was prepared for the disaster and that's why faced worst consequences. This disaster caused more than 1.50000 death and thousands are still missing.

Table 8. Death Toll in all the Tsunami Affected countries.

S. NO	*Country*	*Death+ injury*
1.	Indonesia	94, 100+ 100, 000
2.	Sri Lanka	30, 196+16665
3.	India	15160+ N.A
4.	Thailand	74+ N.A
5.	Maldives	68+299
6.	Malaysia	59+45
7.	Myanmar	2+N.A

Elements at risk from Tsunamis: All structures located within 200m of low-lying coastal area more vulnerable infrastructure facilities in these areas are at risk.

Mitigation

Hazard mapping is an effective tool as a hazard map shows vulnerable areas where Tsunamis can effect, historical data can help showing facts of the past. Evacuation routes should be constructed and mapped on the basis of past experiences.

A mangrove forest management programme should be encouraged like Promotion of community nurseries of mangroves and other species under the coastal bio-shield and agro-forestry programme. Regeneration of fisheries and fostering a sustainable fisheries programme.

Raising artificial coral reefs and managing marine biosphere reserves. Construction of sea walls and dykes.

Reclamation of salinized soils. Development of rules and regulation for coastal ecological security. Promotion of sustainable management of coastal land and water resources (Swaminathan, 2005).

Early Warning Systems

In Rikuzentakata about 1,500 people died according to one estimate, this town was also eradicated by tsunami that hit Japanese with other towns, best defense against any tsunami early warning that allows people to seek higher ground. The Pacific Tsunami Warning System, a coalition of 26 nations headquartered in Hawaii, with a web of seismic equipment and water level gauges to identify tsunamis at sea. Similar systems are proposed to protect coastal areas worldwide.

At the time of emergency people should warn by a warning system. Warnings should be disseminated at local, national as well as international levels People should be evacuated safely from the vulnerable areas.

Community preparedness: community should be prepared to face the disaster and tackle the problems and take suitable actions.

For long term mitigation development plans should take into account the vulnerable areas to reduce the impact of disaster.

Table 9: Largest Tsunamis after 1950.

Country	*Year*	*Cause*	*Deaths*
Chilly	1960	Earthquake	6000
Philippines	1976	"	5000
Indian Ocean	2004	"	230, 210
Java	2006	"	800
Japan	2011	"	18550

Source: Dailyhunt.

Fig. 12: Indian Ocean Tsunami 2004.

Floods

Flood is a temporary inundation of water for a considerable time period floods are the deadliest disaster are maximum loss of life and property is caused by floods in the world. Heavy rainfall and ice melting is the main cause of floods in the flood plains of rivers of the world floods also occur due to collapse of reservoirs and dams, floods also occur due to intensive rains called flash floods. However, areas with no stream or river may also suffer from floods in the case of inadequate drainage- it is generally found in unplanned urban areas, it is called flooding to distinguish from river floods. Streams maintain an equilibrium among slope, sediment load and discharge rate. The rivers are active erosional and depositional agents, the water flows through the river well below or above the bank heights all-round the year. Floods are the integral part of the river and a normal phenomenon. Factors like climate, topography, soil type and vegetation together determine the severity of floods (Subaramanya, 2003). After Bangladesh India is the most flood affected country in the world. The frequency of floods is increasing every year because of climate change distribution of rainfall is highly uneven both in space with long gaps and within short period of time water discharge is very much so that it increase the water level of rivers, hence they flood.

Floods cause heavy damage of life and property in low lying coastal areas because the houses and the properties are washed out. Standing crops and live stocks are damaged.

Types of floods:

1. Flood caused by Rainfall

   Flash floods

   Floods caused by cloud burst

   Seasonal floods

2. **Floods caused by ice melting**

3. Non-precipitation floods

   Estuarine floods

   Coastal floods

   Storm surge caused floods

   Breach Floods

   Floods due to dam burst

Safety precautions

Safety precautions must be taken by people living in flood prone areas.

1. Make an emergency kit which should include a portable radio, torch, batteries, first aid box, dry food items, candles and other essential items.
2. Keep your valuables and documents in a safe place.
3. During flood, move along with your family to safe areas and relief camps for shelter.
    1. Turn off power and gas connections before leaving the home.
    2. Don't enter into the flood water.
    3. Stay away from sewerage live, gutters drains etc.
    4. Don't use electrical appliances.
    5. Use boiled and filtered water.

Mitigation

Floods are common from the time immortal as, civilizations were first proper in the flood plains, they derive both benefits and losses from floods. Floods caused misery and death all the world, they are the deadliest disasters among all the natural disasters, they also effect on the sustainable development of the regions where they occur.

Mitigation measures include both structural and non-structural measures are:

Constitution of reservoirs for collection of monsoon flow to be released in a regulated manner.

Construction of Embankments are mandatory as storage reservoirs in the upper reaches of a rivers do not always in checking floods Embankments obstruct helpful free drainage of land located on outside of the lower reaches of a river. It is essential that while planning the embankments, suitable steps be taken to ensure drainage of the area in the river through sluices.

Embankments require proper maintenance otherwise it proven highly dangerous and may cause great loss. Anti-erosion measures and improvement of flow conditions are also required.

Non-structural mitigation

1. Flood plain zoning and flood proofing and disaster preparedness.
2. Wetland management.
3. Flood forecasting and early warnings.
4. Post disaster relief and evacuation

Some integrated methods of flood management are:

The flood management interventions suggested by WMO are as follows:

Afforestation:

Flora binds the soils and delayed the frequency of runoff hence check the floods and soil erosion both deforestation in the catchment areas of the rivers are the major cause of flooding. The humus content of the soil increases infiltration and also dispose - off water by evapo- transpiration. A thick vegetation cover enhances the absorption capacity of the soils, the effective measure of flood management is therefore to bring areas under dense vegetation and stop deforestation.

Reservoirs and Detention Basins:

Surface water storage by reservoirs, and small check dams are constructed to reduce flood hazards. Storage has to be used in an appropriate combination with other structural and nonstructural measures (WMO, 2007).

Channel Improvement:

River bed deepening, widening and clearing of debris and vegetation increases the water carrying capacity of the river.

Land use planning:

Land use planning is generally adopted in those areas where intensified development on the flood plain is not required.

Development in selected areas can be promoted through incentives. Flood proofing and house building are prove appropriate where development is not intense and properties are scattered Flood proofing of the buildings in flood prone areas can reduce the impacts of floods on the economy.

Flood forecasting

Floods are the disasters that can be forecast through a highly advanced and fully coordinated system, composed of environmental monitoring, forecasts and warning, and their dissemination to the people under threat needs to be established. Meteorologists and hydrologists can help in flood risk reduction up to a great extent by issuing:

Daily weather forecasts and early warnings such as water level forecasts of the river.

Flood forecasts including rates of flow, time of occurrence, duration of the peak flows etc. and Forecast of weather data.

Now great success in flood forecasting has been achieved on many large rivers of the world but on the other hand, warnings of Flash floods associated with heavy thunderstorms in the cities are often very uncertain and unpredictable.

The forecast must be given timely and accurate and surely combined with a community awareness program that trained people how to react and what to do after receiving the warning of floods.

There should be a flood forecasting and warning centre must be established in the community or group of communities that have access to national meteorological and hydrologic data agencies forecasting procedures must be developed and an advance communication system for dissemination of the warnings need to be established. Radars, satellite imageries, wind intensities and expected rainfall amounts all aid in providing ample warning time for flood forecasting.

A unified program needs to be developed that determines the procedures of forecasting, warning, communicating, and evacuating the people. The elements of that unified program are as follows:

1. The Flood monitoring system: Composed of the advance equipment and procedures for rainfall and runoff data recording, analysis of flood and making predictions and risk mapping, predicting flows in areas that have threat of flooding and issuing warnings to institutions and communities to create awareness of flood related problems and make them inform and educated about how to cope with the disaster.

   Emergency plans are prepared which identifies actions to be taken before, during and in post disaster period after a flood management program, for updating, testing and monitoring of flood conditions, and formulation of warnings and emergency plans.

2. Flood warning: Flood warning is way to warn the vulnerable population and make them aware in advance that a flood may occur in the near future at a certain location, actions that are controlled by warnings are the warnings about potential dangers and warnings about imminent dangers that require emergency action.

A warning of a flood event means an existing threat to life and property and it invites responsive action to reduce the danger.

A forecasting and warning system needs to be established, composed of six organizational sub-systems which are as follows:

Forecasting and warning department responsible for collection, evaluation and issuing of warnings, responsible for monitoring the development of a flood threat and for offering instructions and assistance to local emergency organization it also responsible for training of staff.

Emergency organizations responsible for particular activities in their local areas such as door-to-door warning, search and rescue, evacuation of residents, moving valuables, clearing debris, registration of victims, co-ordination with police, firefighters, medical and ambulance services and local utility service providers.

Help can also provide by other organizations such as Red Cross, churches, schools, universities, charity organizations, non-governmental organizations, etc. Media and general public can also help.

The system status reviewed on a regular basis and, if necessary, constantly updated.

Table 10: Deadliest Flood In The World

COUNTRY	*YEAR*	*DEATH TOLL*
CHINA	1975	231, 000
BANGLADESH	1974	100, 000
INDIA	2013	28, 700
CHINA	1980	5, 700
INDIA	1968	6, 200
INDIA, BANGLADESH	1998	4, 892
INDIA	2016	3, 838
PAKISTAN	2010	8, 000
NORTH VIETNAM	**1971**	**1, 600-2, 000**

Flood plain zoning: is done to regulate land use in the flood prone areas to restrict the damage by floods. Through land use zoning we can reduce the losses by putting vulnerable things away from flood prone areas. It places limitation, on unplanned development of both the protected and unprotected areas remote sensing and GiS have great application flood hazard assessment and monitoring.Satellite data with improved resolution is highly applicable for disaster management.

DROUGHT

Drought is a slow onset weather related disaster which is aggravated by human activities. Droughts occur due to variability of rainfall, scarcity of water occurs when drought occur, its effects are devastating, it can start anytime and attain degrees of severity. Its impacts has serious effects on economy as it covers generally large areas, it also adversely affect ecology, biodiversity of a region and reduces food security, during 1967-1991 there were 50 per cent of the 2.8 billion people who suffered from all the natural disaster. Total 3.5 million people were died due to natural disasters out of which 35%

were account for droughts. A drought can last for some days, months and years.

Causes of Drought

1. Rainfall or Precipitation variability

Droughts take place whenever there is scarcity of rainfall and snowfall. When a region remains rainless for long periods, especially for more than a season, then the situation leads to dry conditions and water scarcity which are known as drought.

Sowing done in the expectation of rain and it doesn't rain, drought conditions are experienced In such cases, it is termed as agricultural drought.

2. Human Causes

Human activities can significantly contribute in the management of the_water cycle. Activities such as deforestation, construction, and agriculture negatively impact the water cycle vegetation cover is vital prerequisite for the water cycle as it helps in evaporation, store water, and help in bringing rainfall. In this way deforestation increases evaporation and lessens the ability of the soil to hold water leading to increased susceptibility of desertification. Deforestation can also influence the occurrence of dry conditions since it reduces the evapotranspiration and watershed potential of forests. Activities such as construction and agricultural also reduces the overall supply and quantity of water, results in dry spells.

3. Surface Water Flow reduction

Lakes, rivers, and streams are the primary sources of surface waters in various geographical regions around the globe. In hot and dry seasons or because of overuse of water in certain human activities, these surface water sources start drying out downstream contributing to drought shows that the demand for water is higher than the available water. Irrigation of fields and hydro-electric projects are some of the human activities that can significantly decrease the amount of water flowing downstream to other areas.

4. Global Warming

Burning of fuel by humans contributed to increase in greenhouse gasses into the atmosphere thus resulting in the continued rise of the earth's average temperatures. As a Consequence, evaporation and evapotranspiration levels have risen, and the higher temperatures have led to extreme weather conditions and extended dry spell periods. The global warming situation tends to exacerbate the drought conditions. The worst droughts witnessed in sub-Saharan Africa have been associated with global warming and climate change.

Consequences

The effects of drought are widespread and have devastating impacts on the environment and the society as a whole. Water use is essential for almost every human activity as well as the life of plants and animals. The extended deficiency of water can affect the society in different ways both directly and indirectly. The effects of droughts are thus environmental, social and economic.

1. Drought if prolonged results into serious consequences including environmental, economic and social, low surface and subterranean water levels, lower flow levels, increased pollution of surface water, drying up of water outlets, decreasing bio-diversity and economic losses include lower agriculture, forests and fishing output storage food and drinking water leads to hunger and starvation.
2. Environmental Impacts of Droughts

Animal and plants life is effected in severe droughts. Mainly, the destruction of the wildlife habitats and reduction in water quality and quantity. Many plants and animals completely fail to recover after the drought. The overall climate, the rocks, and soils are also affected, negatively impacting various living and non-living elements of environment.

- **Drying out of water channels and reservoirs**

  Surface water bodies are dry out during prolonged dry conditions which destroy natural habitats. Particularly aquatic life and other wildlife depend on these water bodies die or become vulnerable and the entire food chain is destroyed which damage the whole ecosystem.

- **Soil quality is reduced**

  Moisture content in the soil is essential for soil microbial activities, is reduced in drought conditions. As a result, quality of soil is lowered because of minimized organic activity and continued dry spell which kills soil organisms. As a result the soil is cracked because of dryness and it is led to decertification.

Loss of fertile Land

Drought conditions make it unsuitable for plants and vegetation to survive. Besides, fertile lands are lost as a result of drought, and in consequence, desertification starts. Desertification increases where land become infertile and bare as a result of overgrazing and is exacerbated by drought which makes it difficult for such lands to recover.

- **Reduction in Bio Diversity**

  Animals and other organisms including humans are forced to migrate in drought conditions since they have to move for long distances to get water and food. The prevailing circumstances during droughts make it difficult for the survival of the animals. When the wildlife and animals migrate, they reached in new locations where they can be vulnerable, endangered because of new threats. This leads to loss of biodiversity and disruption of the natural ecosystems.

- **Economic Impacts**

  The economic impacts of droughts are monetary and business losses incurred during droughts by governments, businesses, families, and at the individual level. These are some of the examples of economic effects of droughts.

  During droughts, more money is spend on crop irrigation so as maintain crop yields, lots of water has to be availed for watering the farm animals to ensure the daily water consumption standards are met. Hence, farmers have to spend more money to buy water to keep the crops and livestock nourished with enough water.

- **Crop yields are reduced**

  Often, low crop yields are reduced during drought periods. Therefore, farmers usually undergo major economic losses due to low crop yields. They pay for lots of inputs and labour, but the outcomes are less.

- **Industrial and governmental losses**

  Industries making farm equipment manufacturing and merchandising respectively loss millions of dollars when farmers lack the money to buy their resources. Governments, on the other hand, have to allocate more money and spend even more for drought mitigation as they have to support the farmers and the entire society from the adverse impacts of the droughts. Such governmental monetary expenditure includes funds for emergency supplies, seed funds, and availing other relevant drought mitigation resources.

- **Higher energy costs**

  Prolonged dry spells can reduce the water levels in rivers and dams used to generate hydro-power. As a result costs of energy are higher because the hydro-energy companies are driven to operate below capacity. At the same time, increased energy demands lead to increased cost of grid energy, which leads to economic losses for energy industries.

3. Social Impacts of Droughts

Droughts also have many social implications. they directly effect people and communities.

Some of the implications are as follows:

- **Outbreak of waterborne diseases**

  Shortage of water during drought effects o the quality of water which decreases substantially.

The availability of clean water for drinking and sanitation and cleaning may not be sufficient and High evaporation rate in droughts also increases the concentration level of nutrients, chemicals, and solid particles or impurities in surface waters. Thus outbreak of waterborne diseases such as typhoid and cholera become widespread, especially in poor and developing regions of the world.

- **Hunger, malnutrition, and deaths**

  Hunger, malnutrition and deaths are often witnessed in drought-laden areas. Drought is a key factor for low food production, thus, in poorer regions the effects of malnutrition, hunger, anemia and mortalities are compounded since there is little food available for consumption.

  Lack of sufficient food nutrition is the main factor that directly contributes to diseases and health vulnerability. Hunger, anemia, malnutrition, and deaths are commonly recorded in poorer nations of the world.

Migration

People are migrate to other destinations in search of better living conditions during droughts. This contributed to loss of livelihoods, these are mainly small-scale farmers who are dependent on their farm produce for their livelihood as they are their main or only source of income. The landless labourers, craftsmen, artisans and other people who are dependent on agriculture, forestry and fishing are forced to migrate because they undergo lots of stress, anxiety and are push to indulge in strenuous activities to provide food for their families. Among all Women, children, and the elderly are the most vulnerable.

Drought Preparedness

Mitigation program for drought has three phases:

1. Preparedness and mitigation
2. Prevention
3. Relief

In the preparedness phase first of all drought prone areas are identified, information about land use and land cover, wastelands, flora, soil moisture, amount of rainfall. Monitoring of area through space born aerial cameras and satellites.

Early warnings of drought and forecast

Remote sensing data is most important source of predictions. All the elements of climate (Temperature, pressure, wind, humidity) are required for predictions. Data can be collected from geo- stationary and pole orbiting weather satellites such as INSAT, IRS, SPOT, NOAA, etc. For prediction for drought sea surface temperature, NDVI normalized difference vegetation index, snow covered areas, surface temperatures, altitude, roughness of terrain, soil moisture at surface level and data of water vapour, pressure and temperature.

Monitoring of Droughts

Vegetation index which is derived from satellites shows vegetation stress and it can be used as a substitute measure to continuously monitor the drought conditions timely and such efforts help the policy makers in initiating strategies for loss reduction by changing cropping patterns. NDVI is used for drought monitoring.

Drought Relief

National and state governments of every country are basically responsible for both short term as well as long term relief management, relief of people and animals both is required.

Droughts and its management

Drought management is a complicated process it includes three important components namely,

(1) Assessment of drought intensity and continuous monitoring.

(2) Drought identification and declaration of areas for management of drought.

(3) Development and implementation of drought management strategies.

To ensure effective end result a holistic approach is needed. Currently, all the current knowledge and experience are taken in to account, identify problem areas and strategies are laid down with milestones clearly set through a wide consultative process involving all stakeholders and policy makers. The selection of standard procedures for declaration of drought includes the time of declaration, and the extent of the risk and the vulnerability of various States from droughts can be easily understood and development of optimum procedures for assessment and generation of vulnerability maps can be adopted, to minimize loss of lives, livelihood and property are addressed in an appropriate and symmetrical manner.

Drought proofing of chronically drought-prone areas is mandatory, drought declaration, vulnerability analysis and drought management measures are undertaken on the priority basis. A critical evaluation method is put in place for ensuring that benefits reach the real beneficiaries. Time lag should be reduced by effective use of e-mail, Video Conferencing and mobile phones for effective mitigation.

Remote sensing and GIS techniques and data compilation is promoted to study historical and future trends of the effects of droughts. In the drought intensity assessment, declaration and drought vulnerability assessment institutional participation and use of collective expertise is required for mitigation expert advisory systems can be established for providing advice to the affected population to mitigate droughts, drought assessment and drought declaration are done but not in a proper manner, different organizations did this work in local formats making it unfit for quick analysis and decision making.

The availability of key data used for drought intensity assessment is rarely available particularly in the developing world, for the drought identification data is required on rainfall, reservoir levels. The data collection and its maintenance vary across the world. This leads to unreliable use of Information Technology (IT) tools by various agencies in management of drought. Shortage of Check dams in the rain fed areas results in inadequate storage-water in times of dry spell or drought.

If community participate in drought management activities at the village and Tehsil level, and involvement of Self Help Groups, NGOs and the corporate sector in drought management is also low it reduces the overall value of the effort.

Source: INDIA TODAY.

Fig. 13: India's Deadly drought

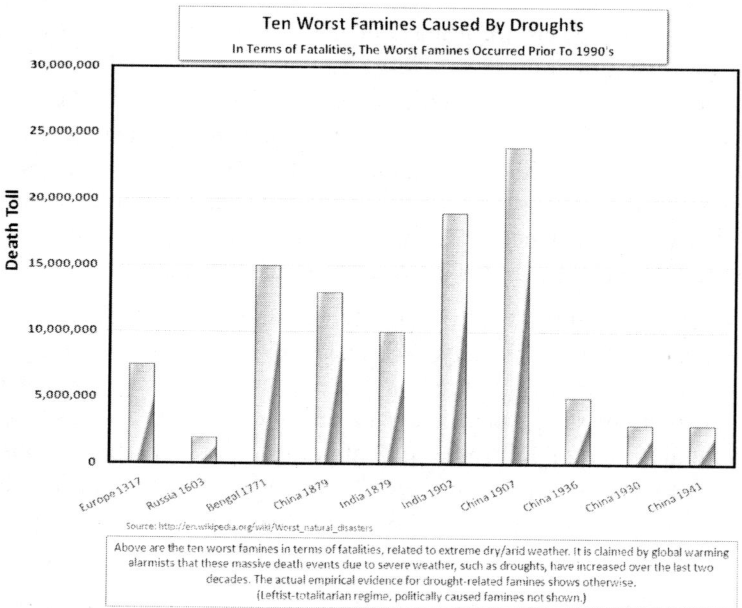

source: Wikipedia

Fig. 13: Ten worst famines caused by droughts.

Chapter - 2

Anthropogenic Disasters

Human activities are also responsible for many disasters which cause serious injuries and deaths as well as property losses, anthropogenic hazards are those hazards which are caused by human actions. They are unlike natural hazards which are caused by natural forces. Anthropogenic hazards may adversely affect humans, other creatures of the world and environment. The frequency and severity of hazards are key elements in disaster management studies. For an example the center of the earth consists of molten material at very high temperatures which would be a severe hazard if contact is made with the core. However, there is no way of making contact with the core, therefore the center of the earth currently poses no hazard.

Mostly the natural disasters are aggravated by anthropogenic activity. Thus natural hazards turn into disasters i.e., severe floods are often caused by deforestation, urbanization, silting of river beds and encroachment in flood plains, destruction of coastal swamps and wet lands is responsible for the severe damage by tropical cyclones and failure to impose building codes and implement earthquake resistant techniques is mainly responsible for the collapse of buildings and human life losses even by low intensity earthquakes.

Human activities are also responsible for the climate change and rising levels as a result of global warming caused by GHGs concentrations in the atmosphere (IPCC, 2007).

Coping with the stress of a disaster can cause psychological as well as physical disorders. Many mental health professionals recommend avoiding overexposure to news broadcasts and to stay away from heavily damaged areas.

Important anthropogenic disasters are oil spills, oil fires, terrorism, nuclear meltdown, chemical accidents etc.

Distinguished features of Man-made disasters

Predictability is mostly nil due to varied causes. Hence warnings and forecasting is not possible, even for the ecological disasters we can't say when a hazard will turn into a disaster.

No system is fully safe, needs timely monitoring and maintenance.

Safety and evacuation plans are either nonexistence or not sufficient. Drills for evacuation are must in vulnerable areas.

Not only the disaster sites and their neighborhood suffer, but the adverse effects are carried farther by winds and waters.

Man- made disasters are increasing in number and magnitude as hazardous industries are growing in number and technological complexity and spread into the countryside. The growing population in general and very large cities with multi million inhabitants increase the vulnerability considerably. The awareness about hazard is in the community is minimal. General public is ill-equipped mentally and physically to deal with disasters and it is generally too late by the time information goes out and outside help arrives.

Response to Man –made Disasters

Man -made disasters have very little scope of forecasts and early warnings, which makes advance preparedness difficult. As accidents such as fires, boat and road accidents, railway accidents, air crash cannot be predicted.

Resource mobilization is essential. Good communication of the requirements of each type of these accidents is considered essential to respond at the earliest to any of the accidents. A perfect co-ordination is required not only within the departments concerned with the accident and with only supporting departments and agencies to deal effectively with the experiences of accidents happened in the past.

The time taken to respond the victims is called Golden Hour is of paramount importance that provide opportunities to undertake timely rescue and relief operations.

In most of the cases disasters occur due to human errors and in a few case malfunctioning of machines is the reason, so awareness and implementation of the safety, and preventive measures, proper maintenance and monitoring is required along with observation of the safety codes is also mandatory.

The rules and regulations must be strictly implemented to reduce the accidents.

The preparedness measures should be (a) source specific, i.e, the source of hazard whether gas leak or radioactive emission etc.

(b)Target specific, i.e., the vulnerability of the site, property and people. Preparedness plans should be made, and practiced for emergencies results from a potential disaster source.

Oil Spills

An oil spill is the release of a liquid petroleum hydrocarbons into the environment due to leakage of oil tankers during its transportation is a form of pollution. Oil is transported from one place to another through ships, thousands of ship sail across the high seas carrying millions of tons of crude oil and other products that are highly polluting. Leaks and spills occur time to time. Ships collide or break oil splits on high seas as a result pose serious pollution problems to the marine life and spoil beauty of beaches To check spills preparedness is required, to deal with them effectively, all possible precautions must be taken to see that such accidents don't occur.

Cleaning up of crude oil from the sea water and beaches is very costly and time consuming. It may take weeks or months it depends upon various factors such as temperature of the water, type of shoreline and beaches of polluted place.

Fig. 14: Brown pelican in Louisiana, 2010. Photo: Getty.

Effect of oil spills in the oceans on the birds.

Oil spill in the sea

Source: Xavier Everaert Sputnik News.

Fig 15: oil spill in the water.

Important oil spills in the past

A number of instances are found in the world of large oil spills some are those which are reported from Alaska, Galapagos, France, Sunder bans, Mumbai and Gulf of Mexico etc.

Oil spills at sea are generally more dangerous than on the land as it is easily avoidable on the land than water.

Oil spills penetrate into the structures of plumage of birds and fur of the animals reduces its insulating ability, and make them more vulnerable to temperature fluctuations.

Oil spills have disastrous consequences for society as it effects on economy, environment and society.

Mitigation There are methods of removing pollutants but none very effective.

Chemical methods are there use of microbes which eat crude oil chemical and mechanical measures can be adopted. The cost of cleaning must be charged from the culprits who do it.

Problems: When oil spills, it starts floating on the water, make a layer on the water and stops sunlight penetration which makes difficult the survival of plants and animals cleaning of spilled oil is a very difficult and costly exercise. Oil spill is very dangerous for aquatic life and cause huge destruction of species.

Oil spills are happening from over a century but interests emerges about three decades back.

Effects of oil spills

1. It reduces aquatic life of the effected water bodies by reducing sunlight and oxygen supply inside the water by blockage of furs of plums of sea birds if they swallow it by chance they died immediately because of its toxic effects. Animals become blind, fish mammals, birds face these often effects many birds and animals starve to death.
2. Economic losses: Huge amount of crude oil lost in the sea is a great loss to the country whom it belongs to, large cleaning

costs is involved the responsible company has to clear it great help of government is also required. The people involved in cleaning, face a number of health problems later in life. Their medical treatment is the responsibility of the government. All this cause heavy burden on the economy.

3. Effect on recreation activity, the tourism activity suffers a huge set back as most of the tourists stay away due to pollution, activities like sailing, swimming, rafting, fishing, gliding cannot be preferred.
4. If oil reaches the shoreline or near the coast it reacts with the beach sand and rocks, vegetation and territorial habitats of animal and humans caused erosion and contamination. Waves, currents and winds move the oil from one place to another. Beach sand and gravel when saturated with oil may become unable to protect natural vegetation and population of that region. Rocks and boulders coated with oil become toxic to coastal organisms.

Oil spills can cause enormous harm to deep oceans and coastal fishing the immediate effect may be mass mortality of fish and other organisms. Intercept the food chain of the effected waters.

Recovery: Depends on type of water body involve such as open ocean, coastal or wet land, the pollution clearance measure includes skimming, filtering, dispersal of oil sheets into small droplets to increase sunlight and oxygen penetration, large sized birds and mammals are sometime rescued and cleaned, but micro sized organism are generally ignored.

Prevention

Prevention of oil spills is of utmost importance. The oil prevention Act was enacted in 1990 by the U.S congress to prevent the oil spills also for planning, response and recovery.

Table 11: Largest oil spills.

Tanker	*Location*	*Date*	*Tons of crude oil (Thousand)*
Kuwait oil fires	Kuwait	Jan16-Nov6, 1991	136000
Kuwait oil Lakes	Kuwait	Jan16-Nov6, 1991	3409-6818
Lake Gusher	United states, California	14 March1910-sept.1911	1200
Gulf war oil spill	Iraq, Persian gulf	Jan19-Jan 28 1991	818-1077
Deep water horizon	United States Gulf of Mexico	April2010-july2010	560-480
Ixtoc	Maxico Gulf of Mexico	Jun1979-march1980	287
Atlantic oil spill	Trinidad	19 July 1979	285
Fargana Valley	Uzbekistan	2 March 1992	260
Nawruz fields	Iran Persian Gulf	4 Feb..1980	260

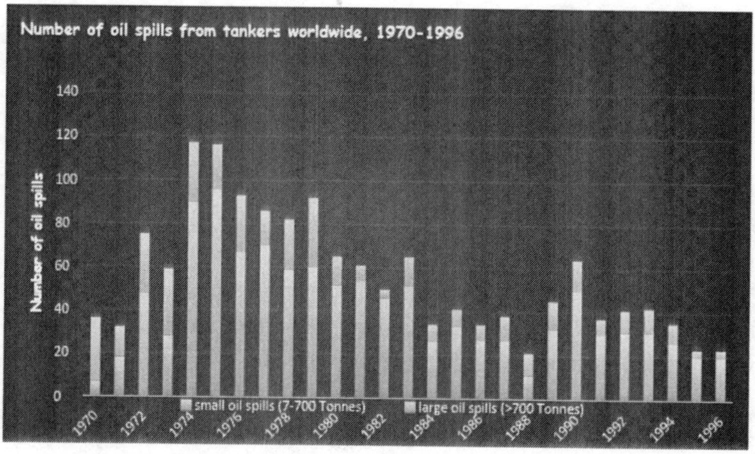

Fig. 16: Number of oil spills from tankers worldwide, 1970-1996.

Fig. 17: Oil Fires.

An oil fire is always dangerous. These occur in a flash and last for several days in the wells which produce crude oil. Immediate safety measures to avoid a flash fire otherwise great loss of production could occur. Emergency evacuation of personnel and immediate fire quenching measures are required. Isolate the fire immediately and alert everyone in the area should be warned when a fire starts. A regular fire drill should be held, surprisingly to check the alertness of the people to deal with any flash fire. Fire extinguishers should be fitted everywhere they are required and there should be some trained people who can handle it perfectly as there is no time for trial and error method while tackling with fire. Services of special teams from Texas, USA are used around the world they are expert in quenching fires they help when local efforts are fail. This is a specialized field, they have specialized equipment and skills to deal with the situation.

The Gulf oil fires

According to one estimates the Gulf fires created losses of at least $ 22 billion and caused damage to physical infrastructure, like pipelines and wellheads, oil was wasted, and environmental destruction.

The effects of the oil fires mentioned above along with the great fire the disaster also negatively impacted the physical health of people, including troops, fire-fighters, and civilians.

The burning oilfields released hazardous pollutants like sulphur dioxide, carbon monoxide, and lead. These pollutants negatively affect human health and are particularly injurious to those who are already suffering from respiratory diseases.

Various studies have described the ill effects of the burning Kuwaiti oilfields on health. According to one study the U.S. soldiers stationed in Kuwait frequently reported problems of coughs and f breathing. Citizens of Kuwait were also impacted by the oilfields fires, there respiratory tract infections were increase. This trend coincided with the increase of fires.

For those who exposed to the oil fires for only a short while the event did not permanently affect their health in a negative way. Exposure over a long period, however may have more serious consequences, such as cancer. Studies on the long-term effects of exposure have not conducted or not fully known.

Terrorism

Terrorism can be defined as the systematic use of_violence to create a general climate of fear in a population to full fill a particular political objective. Terrorism has been practiced by political organizations with different objectives such as rightist and leftist objectives, by nationalist and religious groups and by revolutionaries. Definitions of terrorism are usually complex and controversial, and, because of the inherent ferocity and violence of terrorism, the term in its popular usage has developed an intense stigma.

It was first coined in the 1790s to refer to the terrorist activities used during the French Revolution by one group against their enemies. The Jacobin party of Maximillian Robespierre carried out terror involving mass executions by the guillotine, although terrorism in this usage implies an act of violence by a state against its domestic enemies, since the 20th century the term has been applied most frequently to violence aimed, either directly or indirectly.

Since the 20th century, ideology and political opportunism have led a number of countries to engage in terrorism, often for supporting of movements of national liberation. Hence, it is said that one man's terrorist is another man's freedom fighter. The distinction between terrorism and other forms of political violence became blurred particularly because many groups often used terrorist tricks and legal issues were similarly many groups often employed terrorist tactics and issues of Jurisdiction and legality were similarly hide out.

The 20th century witnessed great changes in the use and practice of terrorism it became the mark of a number of political movements stretching from the extreme right to the extreme left of the political lobby, technological advances such as automatic weapons and compact, electrically detonated explosives, gave terrorists a new mobility and inevitability, and the growth of air transport provided new methods and opportunities. Terrorism was virtually an official policy in states such as those of Nazi Germany under Adolf Hitler and the Soviet Union under Stalin. In these states arrest, imprisonment, torture, and execution were carried out without legal guidance or restraints to create a climate of fear and to encourage adherence to the national ideology and the declared economic, social, and political goals of the state.

Terrorism is not a recent phenomenon it is as old as human history. There are many examples all over the world of terrorism. It is very difficult to define terrorism, it may be defined as unlawful act of violence against the people or property. The terrorist threatens death or destruction as a part of his program of action.

Factors responsible for terrorism

Misery, Frustration, religious fundamentalism, political intolerance, insecurity are some of the factors of terrorism which plunges the society into chaos (Hindwan, 2001).

Prevention

Awareness and counselling of masses can help in reducing this act, governments of the countries should prepare with all the related equipment. People must be educated about the past terrorist attack and play an important role in confronting with the problem.

In response to this threat, gradually over five decades of work, the international community has developed a common universal legal framework against terrorism. This framework is comprised of the19 Universal legal instruments against terrorism along with the relevant United Nations Security Council Resolutions. The implementation of these conventions, protocols and resolutions is informed by the guidance provided by the UN Global Counter Terrorism Strategy along with United Nations General Assembly Resolutions.

The Terrorism Prevention Branch (TPB) of the United Nations Office on Drugs and Crime (UNODC) has a specific role to play in these international efforts. For over a decade, TPB has been the key United Nations entity providing legal counter-terrorism technical assistance to Member States. As mandated by the United Nations General Assembly, TPB works to assist Member States, upon request, with the ratification, legislative incorporation and implementation of the universal legal framework against terrorism.

Nuclear Reactor Meltdown

A nuclear meltdown is an accident of serious nature that occur because of overheating. It is an accidental melting of a nuclear reactor. It happens when heat generated by a nuclear reactor is more than the heat removed by the cooling systems.

Nuclear accidents are common on the worldwide basis the first one was Chernobyl disaster in 1986, is proven that technical innovations cannot eliminate the risk of human error in nuclear plant accidents from 1952-2009, these accidents cause loss of human life and property..

1. In 1979 Three Mile Island accident in Pennsylvania was occur due to a series of failures in security system at the reactor.

2. Chernobyl disaster in Ukraine is the world's largest nuclear disaster happened in 1986 due to rupture of the reactor vessel and a series of steam explosion that destroyed reactor.

3. Fukushima nuclear meltdown occur on the 11 march when a powerful earthquake with a magnitude of 9.0 rocked in the Northern Japan, people are asked to evacuate living in 30 km radius. It was a level 7 nuclear accident that is the highest level on the international nuclear event scale (INES) which has reached only by Chernobyl.

Effects

The effects of nuclear accidents are of severe nature it can cause radiation sickness, mainly it harms to workers and emergency teams in a stricken plant it causes acute radiation syndrome like skin rashes, vomiting and dirihoea, and even coma and death can occur.

Radiation damage DNA, the tissues which contain many cells are most at damage risk.

Major health effects of nuclear disaster

1. Fires and explosions takes place when a nuclear meltdown occur sometimes the fires ignited by the explosions are aggravated by storms prevent the escape of survivors, they are also a cause of casualties.
2. Initial radiations large amount of reduction and gamma radiation are released can cause fatalities and death.
3. Disaster at the Chernobyl nuclear power plant in April 1986 in Ukraine was the product of a flawed Soviet reactor design coupled with serious mistakes made by the plant operators. It was a direct consequence of Cold War isolation and the resulting lack of safety measures.

Chernobyl Tragedy

The accident destroyed the Chernobyl reactor, killed 30 operators and firemen within three months and several further deaths later on. One person was killed immediately and a second died in hospital soon after as a result of injuries he received. Acute radiation syndrome (ARS) was originally diagnosed in 237 people onsite and involved with the clean-up and it was later confirmed in 134 cases. Out of these, 28 people died as a result of ARS within a few weeks of the accident. Nineteen more location of Chernobyl.

Source: Daily Mail.

Fig. 18: Chernobyl nuclear reactor.

Subsequently died between 1987 and 2004 but their deaths were due to radiation it is not confirmed. Nobody offsite suffered from acute radiation effects although a large proportion of the childhood thyroid cancers are diagnosed since the accident, are may be due to be due to the intake of radioactive iodized water. Large areas of Belarus, Ukraine, and Russia and beyond were also contaminated in varying degrees.

The Chernobyl disaster was a unique event and the only accident in the history of commercial nuclear power where radiation-related fatalities occurred. The design of the reactor is unique and in that respect the accident is thus of little relevance to the rest of the nuclear industry.

Outside the then Eastern Bloc. However, it led to major changes in safety culture and in industry cooperation, particularly between East and West before the end of the Soviet Union.

Chemical Disasters

Chemical disasters occur due to release of poison chemicals and gases from the containers because of mishandling in their storage. These accidents also cause due to explosion at the production factory of hazardous material, an accident can cause in the transportation of hazardous material. Failures in the safety systems of chemical plants. Due to a natural disasters such as earthquakes, cyclones, etc. chemical disasters also occur through damage and destruction to chemical units where chemicals are stored.Failures in safety systems of chemical plants.

Chemical terrorism

Chemical disaster can also be caused due to unfair use of chemical warfare agents by terrorists. These chemicals are Sarin, Chlorine, Sulfur Mustards, Hydrogen, Cyanide etc.

Impacts

Chemical disasters lead to serious and varied impacts. They can cause explosions, and fires. The most hazardous effect of chemical disaster is the pollution of air water and food chain disruption due to ecological damages. Many health impairments can even extend to coming generations.A chemical disaster can cause the following consequences.

Physical Damage

This includes damage or destruction of infrastructure and buildings. A chemical accident during transportation can damage to the means of transport viz. vehicle, rails etc.

Deaths

Chemical disaster may result in large scale causalities, immediate disposal of dead bodies also needs planning.

Environmental Damage

Chemical disasters destroy environment by contaminating air, water, land, crops, vegetation and animal life. Many birds and animals and other organisms either died or migrate from the place thereby creating an imbalance in the eco-system.

Management

In India, the ministry of Environment and Forests is the nodal agency for the management of chemical disasters, the ministry list out all the hazardous units of the country.

The Department of Environment of the ministry and controller of explosives are jointly responsible for setting and monitoring of the safety standards.

The Chemical Accidents Rules of 1996 is outlined the procedures of Chemical Disaster Management. A layered structure is made to control such disasters at central, State and District levels. As per law, all industrial accidents have to be reported to the Director General Factory Advice Services and Labour Institute.

Preparedness and response

- **Hazard Mapping**

  Accurate location of type and quantities of hazardous material production and utility sites and store houses should be located on the map at district and state levels.

- **Identification of hazardous material**

  Hazardous Chemicals should be stored and handled carefully they must be divided into categories on the basis of the danger extra care should be paid to their packaging and transportation.

- **Inspection of chemical plants and store houses**

  A regular inspection of these places is required to avoid accidents. Most of the chemical disasters can be avoided by laying down norms strictly.

- **Monitoring of Toxic Waste disposal**

  Disposal of waste by the chemical industries should be monitored regularly, otherwise it will pollute the soil and underground water of dumping places, dumping of waste in water bodies is also very dangerous.

- **Monitoring of Pollution levels**

  Continuous monitoring of environment must be done to check the pollution level and warn the authorities about the possible danger.

  Emergency management plans need to be prepared for on - site and off-site contingencies.

A chemical disaster management plan should include the following aspects:

Awareness among the workers and people living in the premises of the plant. Every plant site should have some kind of warning system for emergencies like alarms etc.

Plan for evacuation and provision of medical aid during and after the disaster. Firefighting arrangements and communication system must be there.

Practical drills, list of specialists and organizations who could provide assistance should be maintained.

Building Fire

Fires in the buildings are disasters of serious response as fire loss is a national loss as what burns never returns. Among all the hazards fire related accidents have high degree of risk and create a great problem some times. Fires always cause heavy loss of life, property, and injury both in private and government spheres.

The fire accidents are increasing mainly due to lack of fire safety norms in public and private places.

Some historical fires are:-

Great Bombay Dock Fire (1944) when 336 persons died and 1040 people were injured including 66 fire service men, the fire broken out on S.S.Forte Strikene, which was carrying 1200tonnes of explosives.

Uphaar Theatre Fire Delhi (1997)-During the screening of a film in the cinema hall a fire started at 5P.M. in the evening. The fire was started from a transformer in basement. In this incident 57 people dead and 105 injured.

Arson Fire at Hyderabad (1984) Fires raged for three days and a large number of fire engines were deployed.

Fire in Hindustan Building (1980) 17 floor high building caught fire. Extensive damage occurred on and above 8th floor. 22 fire tenders and 2 Hydraulic platforms were employed to extinguish.

Bombay Asha Mill Area fires (2017) in a recent episode 14 people including 11 ladies were died in a public place where restraunts and pubs were closely located due to unsatisfactory safety norms.

Mitigation

For dealing with fire accidents, we must be able to understand the phenomenon of burning.

Carbon Monoxide (CO) is produced by improper combustion. CO can kill before one is aware of its presence. If CO poisoning is suspected it is advisable to evacuate all occupants immediately, turn off all appliances and sources of fire and ventilate the house by opening doors and windows.

Smoke and toxic gases kill more people than flames do, it makes people drowsy, disoriented and short of breath.

The fires on the buildings could develop into major disasters. Good housekeeping must be ensured. Faulty electrical appliances should be repaired or replaced immediately. Keep means of escape clear of obstruction. Fire rescue drills should be carried out at regular intervals.

In case of fire do not use lifts, elevators use stairs.

Biological Disasters

Introduction

Biological disaster has coexisted with human society since the time immemorial. With rapid advancement in medical sciences and prevention and social medicine, the impact and frequency of such disaster have reduced to great extent in advanced countries. But the poor and developing countries continue to suffer due to biological disaster. It is, therefore, important to understand the measures of managing biological disaster and mitigating their impact on communities.

What is a Biological Disaster?

A biological disaster is the disaster, which causes sickness and fatalities in human beings and animals at mass scale, when they come in contact with biological hazards in the form of living organisms, such as, bacterial, Virus, fungi, etc. Destruction of crops and plantation also falls within the ambit of biological disaster.

All communicable diseases, either of human beings or livestock are potential biological hazard. They spread widely, affect large number of people in communities, sometimes across the geographical limits of provinces and nations.

Biological Disaster have caused havoc in human settlements in the form of communicable disease since the ancient times. Plague savaged Europe for 300 years from 1300s to 1600s. About 20-25 million or about one-third of the population was eliminated by the deadly epidemic across the continent. Plague killed 200 people at Surat in Gujarat in 1994.

The pandemic influenza outbreak caused 20 million deaths across the world in 1918-1919. Similarly small pox, Ebola, and yellow fever have been causing havoc throughout the world, mainly in underdeveloped and poor societies. The small pox has been eradicated with its virus. It is preserved only by some advanced countries for research purposes.

Biological disaster essentially appear in the form of epidemics or pandemics, which are caused by microorganisms. Different microorganisms cause different types of communicable diseases. The micro-organisms, which cause communicable disease could be categorized as follows:

- **Bacteria-** These are small living organisms. They can be grown on solid or liquid culture media. The disease caused by bacteria is usually treatable with specific antibiotic therapy.
- **Virus-** These microorganisms share replicate in living cells and cause disease, which are mostly non-responsive to antibiotics. Such disease may sometimes respond to antiviral compounds.
- **Rickettsia-** These microorganisms share characteristics of bacteria and virus. In the case of virus, they grow only within living cells; and in case of bacteria. Besides, they use oxygen and are susceptible to antibiotics.
- **Fungi-** These are primitive plants, which draw nutrition from decaying vegetable matter, generally fungi form spores and free living forms are found in soil, fungal disease normally responds to anti-microbial drugs.
- **Toxins-** are poisonous substances produced by living plants, animals and microorganisms. Some toxins can be produced by chemical means also.

Classification of Communicable Diseases

1. Diseases transmitted by contact

   Conjunctivitis

   Mycosis

   Scabies

2. Sexually Transmitted diseases

   Gonorrhea

   Syphilis

   AIDS

3. Vector born diseases

   Malaria

   Dengue

   Yellow fever

   Recurrent fevers

4. Diseases transmitted through fecal matter

   Diarrhea

   Cholera

   Hepatitis

   Typhoid fever

   Hookworm disease

5. Diseases spread though air

   Acute respiratory infections

   Tuberculosis

   Measles

   Meningitis

   Whooping cough

Globally efforts were made to eradicate communicable diseases. Smallpox is among the communicable diseases totally eradicated so far. But after the eradication of smallpox thirty new pathogens have been identified including AIDS, Ebola virus, lassa Virus etc.

Factors Contributing To Vulnerability

Urban, semi-urban and rural population all are vulnerable to biological disaster, through for different reasons and in varying degree. Some of the factors uniformly applicable are:

- **Population growth**

  Leading to sub-standard and unhygienic living conditions, presenting a perfect condition for epidemic to outbreak.

Poverty

A logical consequence of over population limits the capacity of individuals and communities to limit or nullify the impact of epidemics.

- **Lack of Rapid Response and Epidemic Control**

  Paucity of medical resources coupled with geographical locations and problems of communication make communities in rural areas comparatively more vulnerable.

- **Low Public Awareness**

  Lack of basic health and hygiene education and in some cases superstitions add to vulnerability of certain sections of population.

- **Poor Health and Malnutrition**

  Poor health and malnutrition lead to depleted body resistance to diseases. Thus certain groups in urban areas, and woman and children in backward rural areas becomes more vulnerable.

- **Poor State of Health Care System**

  Care less approach to public health and safety coupled with meagre resources at the disposal of civic bodies at all levels also contribute to enhancement of vulnerability to biological disasters.

- **Congestion in Urban Areas**

  Congestion in urban areas leads to problem of waste disposal, which provides fertile ground for various diseases to spread.

- **Bio Terrorism**

  Ignorance towards emerging threats of bio-terrorism, in general, enhances the vulnerability.

- **Modern Means of Transport and Communication**

  It is a paradox that modern means of transport and communication have made the world shrink, which also vulnerability to communicable diseases because of frequent travel and greater social mixing.

Adverse effects of biological disaster

This disaster is increased the vulnerability due to economic consequences and levels of poverty.

Post traumatic disorders are common in the patients and their families.

Crisis of labour who migrate in search of employment. Loss of livelihood due to economic losses.

Preparedness and mitigation

To mitigate biological disasters there must be sufficient medical facilities, well trained medical professionals availability of vaccines and drugs. Adequate stock of diagnostic kits should be available at cheaper rates.

Chapter - 3

Disaster Management

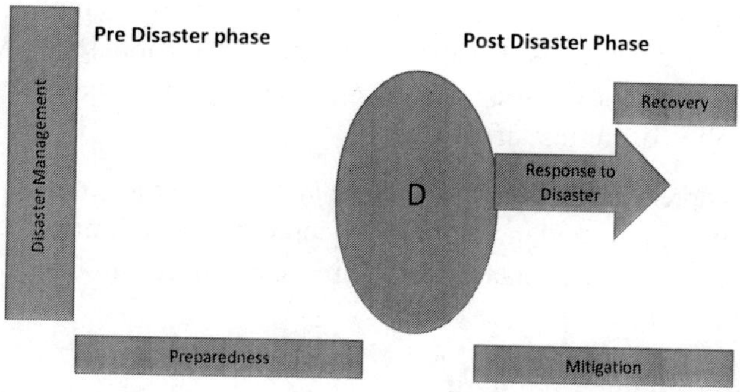

Fig. 19: Disaster Management.

Disaster = Hazard + Vulnerability capacity

Disaster Management – Primarily focusses on the preparedness response and recovery.

Disaster management is an act which includes all the aspects of planning and response to disaster both pre and post emergency periods. Disaster management was developed to reduce the damage and losses incurred by disasters to both life and property.

It involves the following –

(1) Proactive disaster planning to reduce risks.

(2) Minimizing loss through effective preparedness and response.

(3) It creates more effective and durable recovery.

In the present scenario there is greater awareness than before for disaster management mainly due to some recent disasters – Indian Ocean Tsunami in 2004, oil spills, nuclear reactor failure in japan after an earthquake, flash floods in Uttrakhand region of India, Nepal earthquake etc.

The Indian ocean Tsunami in 2004, proved to be a wakeup call for the governments of the countries including India which were adversely affected by it due to lack of preparedness for such a large scale disaster. The impact of the disaster was truly on a huge scale as the United Nations realized that every country requires a proper disaster management strategy which is fully regulated by its government by providing a framework fully supported through financial and administrative measures. The disaster framework is supposed to work round the clock, ever prepared to provide assistance considering disaster often occur unalarmed.

A comprehensive disaster management strategy requires expertise, education, training and motivation of people. It is a multidisciplinary approach as it requires planner, economists, engineers, architects and scientists.

Institutional arrangements

Policy development is required at national and local levels to put together a comprehensive and well-coordinated disaster management strategy that makes optimum utilization of available resources, needs a shared disaster management policy. Without it prevention, preparedness and response will be fragmented and ill coordinated leading to ineffectiveness (WHO 1999). Development of disaster management policies requires an active procedure of analysis, consultation and negotiation among government and non-government entities.

Hazard Mapping

It is a very important tool which identifies risk and plays a very important role in disaster management. Historical records, physical data and computer simulations help us to produce detailed maps providing information about risk zones of landslides, floods, earthquakes, volcanic eruptions, avalanches and industrial accidents.

We can use the maps of areas surrounding the hazardous regions along with, data of wind direction and velocity to predict the scale of possible hazards and determine the method of evacuation or other emergency responses if leaks or explosions occur. For the identification of physical hazard in a region both Public and private records of past industrial activity may prove valuable.

Maps are prepared according to different hazards, these maps are useful for various purposes.

Flood hazards are mapped in it we can superimpose different return periods of flood in different extents of the flood plain, the scale of maps appropriate for hazards maps depends both on the use and the amount of data available. Data of spatial distribution of different hazards like earthquakes, landslides, cyclones and floods is now available through various means so that we can now easily prepare hazard zonation maps on the basis of variation of risk over a region. Such micro-zonation maps are very important for land use planning. Micro-zoning maps can be used for a single event of a single hazard, multiple events of single hazard and to show the combine impact of several hazards.

Settlement areas that are at risk from various hazards are indicated by the hazard maps. Separate maps are prepared for each hazard. These maps can be used to delineate the areas that could be going to flood by the next storm and possible locations where avalanches could be prevalent and slopes where a landslide could occur.

The hazard maps also provide detailed information about the causes and spatial distribution, intensity and probability of occurrence of a natural hazard.

The hazard maps of Delhi were brought out, providing area-wise details of the risks that different parts of the city could face in the event of an earthquake. These were prepared by Earthquake a risk Evaluation Centre under the India Metrological Department. The new maps would not only be useful for new constructions, the existing buildings which had been built when the entire city was considered as one homogeneous region, could also be retrofitted to make them safe. (The Hindu, 2005).

An interesting development in U.S and other developed countries is the introduction of "Shake Maps". Which can be prepared in few minutes after an earthquake and sent to emergency response teams. Such maps can be used for response, loss assessment and for public information.(Mishra, 2002)

Uses of Hazard Maps

1. **Spatial planning**: hazard maps are vital for the regional planning because they provide a basis for spatial planning processes such as identification of hazardous zones in development plans and formulation of building codes.
2. **Risk reduction measures:** These maps help in implementation of the risk reduction measures and formulation of prevention measures e.g. flood protection structures and avalanche barriers etc.
3. **Instruments used in emergency planning:** Hazard maps indicate areas at risk and the events that are most likely to happen, this information can be a useful source of orientation in emergency planning.
4. **Create awareness among the population:** By demonstrating potential risks to the population, hazard maps assist for the awareness of protective measures.

Importance for landowners

These maps help in the planning of structural protective measures in a risk area, they are often available in the relevant communal authority.

Hazard index maps are also prepared, these map provides a general overview of the hazard situation on a scale of between 1:10'000 and 1:50'000. It is based on geo scientific data and simple model calculation.

Hazard maps are prepared on a scale of between 1:4'000 and 1:10'000 and thus give a detailed overview of the hazard situation. The level of detail provided by hazard map is more extensive than that is given by hazard index maps. They provide complete information about the causes, spatial distribution, intensity and probability of occurrence of possible events going to occur.

Functions of hazard maps

The main functions of a hazard map are to identify the risk and to make it known to residents. The hazard maps cannot stop a disastrous phenomenon, but the effective use of hazard maps can decrease the extent of damages.

Objectives of hazard maps

The main objective of hazard maps is to provide the information to the people on the range of possible losses and the disaster prevention measures to reduce the impact of disasters. It is provided to the residents with understandable clear information.

Types of hazard map

The types of hazard maps are as follows:

(1) **Hazard maps for the residents:** This type of map has main objective to provide information to the people living within the risky areas having danger of damages, the details of dangerous and safe places, and the basic knowledge on disaster prevention are given to the residents. Therefore, it is essential that such information is represented in an understandable form.

(2) **Administrative maps**: These maps are basic tools for administrative agencies to plan for disaster prevention activities. These maps can be successfully used for the

establishment of a warning system and for the evacuation of population, also for land use regulations we can use it. They may also be used in preventive works.

Impacts of hazard map

All hazard maps are not acceptable to residents, landowners and land developers can also have fear of the fall of land prices so they oppose to the release of maps for the public. In addition, the top officials in the government may object to the release of the maps, as they have fear of the criticism of people that the government is careless about the disaster preventive works even though they have warnings of occurrence of disaster.

Hazard maps and disaster prevention methods

(1) A hazard maps is an important mean for disaster prevention, however, generation of a hazard map is not the final goal, after the completion of a hazard map, and these maps can be used to provide information about the hazardous areas for residents and to help them response to warnings and evacuating measures.

(2) A hazard map will be useful only when it is publicly available to residents. The administration should make efforts to educate residents by providing disaster prevention information to them by means of hazard maps. The community can make inform through school education, media and awareness generation programmes and trainings at least once a year or more and repeatedly disseminate the disaster prevention information through various methods. Mock drills can be plan to increase the resident's perception.

(3) We can use a hazard map for the economic evaluation of a preventive work and for cost assessment, as a result, the priority of preventive works can be determined.

(4) As described earlier the hazard maps make extensive use of previously available empirical data of past disasters, such a data is extremely important as it gives a better understanding of hazards and also to collect the data that is statistically

significant to estimate the scale of a phenomenon. This helps in estimating the scale of phenomenon and to determine the criteria for forecasting and warning of disasters.

Vulnerability assessment

It is done to identify hazards and their possible impacts on a community if they occur. This is an early strategy in disaster management.

Vulnerability assessment provides us vulnerable areas where we can improve ourselves through preparedness and mitigation.

According to Carter (1991) the national disaster management should provide a comprehensive approach to disaster management, serve as a nodal point for department activities, provide a basis for co-ordination among different agencies, and provide a setup for review and evaluation.

Monitoring of vulnerabilities

Monitoring of vulnerable areas is very important, in this way we always are informed about what is going on and accordingly stay prepared. Monitoring of fault lines, rivers, cyclones, etc. is done for this purpose.

Disaster management has three phases

(1) Pre-disaster phase

(2) During disaster

(3) Post disaster

Pre-disaster Phase

* Prevention

* Mitigation

* Preparedness

In this phase identification of risk and vulnerability is carried out and corrective steps can be taken for prevention and mitigation.

Fig: 20 phases of Disaster Management

Long term measures like improvement of water supply to reduce the health risks to people so they can be physically fit to face numerous challenge, construction of levees to prevent flooding and water shed management to prevent droughts, afforestation to prevent landslides, construction of earthquake resistant buildings to prevent collapse by enforcement of construction guidelines and zoning regulations, creating awareness and giving people knowledge about such measures to prevent and mitigate disasters s which are faced in the aftermath of a disaster.

Disaster Prevention includes those activities which are carried out to prevent a natural phenomenon or hazard from having harmful effects on people and their economic resources. An ignorance and delay in preventive action could completely destroy the lives, economy and the resources within a region.

For developing nations, prevention is perhaps the most important component in managing disasters, however, it is clearly one of the most difficult aspect to promote at the same time. Prevention planning is based on two indicators:

1. **Hazard Identification:** Identification of hazard is the most important component, It includes Identification of the threats

facing a community and vulnerability assessment or evaluation of risk and resilience of a community to handle the consequences of the disaster. On this basis the disaster managers decide the prevention strategies. Disaster prevention includes measures which can eliminate the root causes that make people vulnerable to disasters.

2. **Preparedness:** The measures that ensure the organized mobilization of personnel, funds, and equipment's, and supplies within a safe environment for effective relief are paramount. Disaster preparedness is building up of capacities before a disaster situation manifests in order to reduce its impacts. Measures of preparedness include inter alia, availability of food reserve, emergency reserve fund, seed reserve, health facilities, warning systems, logistical infrastructure, relief manual, and shelves of projects.

Preparedness and planning

Disaster preparedness planning involves identification of organizational resources, determination of roles and responsibilities, development of policies and procedures and planning activities in order to achieve a level of preparedness to respond timely and effectively to a disaster.

The actual planning process is preliminary in nature and is performed in a state of uncertainty until an actual emergency or disaster occurs. After a disaster occurs, plans must be adapted to the actual situation. Planning priorities are disaster specific varies according to emergency at which the plan is developed.

Preparedness plans must be prepared in consultation and cooperation with those people who approve them and implement the plans. Every Nation should have appropriate level of preparedness and a corresponding level of planning. Agencies with different resource levels can choose the planning elements that best suit their objectives. At the community level, Red Cross/Red Crescent Societies can play an important role in supporting the organization of disaster planning committees who can be responsible for developing a community disaster preparedness plan.

The need for preparedness planning

The concept of preparedness planning is very crucial for those involved in disaster management during an actual emergency, quick and effective action is required; however, this action must be depend on proper planning in place before a disaster strikes. Appropriate actions taken to response the disaster play an important role in the saving of lives and resources. Preparedness Plan must include determination of evacuation routes, identify emergency water sources, identification of command and communication procedures, and training of response personnel and educate people about hot to move in case of an emergency, all of these measures will be frame on the basis of minimum time limit, and effectiveness of the response measures.

Because of its future-oriented nature, preparedness planning is more difficult than planning an emergency operation in response to an actual disaster.

In a country like India where resources are scarce a future planning for a disaster which may not occur is not easy as it use the resources which are already limited. Preparedness planning must be an ongoing process, even at times of relative normality. All the development plans and updating of a plan is done after a major disaster has occurred either in the country or in another part of the world and, try to find out the response measures were slow or ineffective resulting in higher deaths and suffering.

The planning process

It is a continuing process not a one-time event. The plan may have detail specific objectives and preventive actions, the objectives need to be corrected and refined during an actual emergency. Planning may be ineffective if all stakeholders are not included in the process. Especially those who are responsible for implementing preparedness or emergency activities are more likely to comply if they feel that their views are incorporated into the planning process. Plans should be created by a person who is known to region for which planning is done as experiences shows that if external person is involved in it the results are erroneous or not useful. Therefore, a perfect team is desirable which is from the local area or they must know the problems prevalent in the region for whom they are planning.

During the communication and coordination stage of plans for setting and updating of specific aims and objectives, national Societies should communicate with the government agencies and other non-governmental organizations working in disaster response. This will improve planning. Through direct coordination, responsibility for different operations and plan can be divided their actions accordingly.

Representatives of various agencies working in various areas, such as health, shelter and relief distribution. Ministries and departments for Civil defense, Fire brigades, Health departments and ministries, International agencies and local authorities also work in disaster response.

The above organizations play their respective roles and responsibilities in disaster preparedness and response and make the necessary communication and coordination mechanisms at different levels. Any plan formulated must be discussed widely among the various organizations and stakeholders, planners should encourage viewing, criticism and discussion from those who must approve it and implement its components for this it is mandatory that the plan must be communicated to headquarters and departments, branches, external agencies and international NGO may all be providing first aid delivery, shelter and food. In this case coordination of activities is required to avoid unnecessary duplication of services. A preparedness plan should maintain a list of names and responsibilities during emergency, and contact numbers and addresses for the emergency response focal point, the team members at each operational level and people in charge must have all the information with them.

Management of administrative activities for a preparedness plan required that each agency should also identify the activities for them and their anticipated level of involvement in the event of an emergency.

To determine the activities before, during and after the emergency a preliminary list of activities can be made. This assessment can be made on the basis local capacity and the extent to which additional resources will be required to respond.

The operational procedures are also determine existing priorities should also be reviewed. Provision of public participation in the planning process, is important for self-reliance and setting up of goals to provide a stable management environment during an emergency situation.

Develop an early warning systems which is an extremely important component of preparedness, prevention and mitigation, it is essential to develop and use early warning systems warn the vulnerable population in a hazard threaten region.

Early warnings if issued in time losses can be reduced up to a great extent for example, after early alert of cyclone Philaen in India, enabled evacuation of 4.4 lakh people before it reached the region and as a result a lot of damage life and property was prevented. If proper early warning systems were put in place earlier in 2004, then the tsunami of Indian Ocean would have not been so deadly as it was, this was due to lack of awareness and preparedness. Thus early warning systems are very important in pre-disaster period as they warn us beforehand and we can do timely preparations to deal with disasters. Today through advanced technological methods of remote sensing and GIS we can achieve much more accurate and reliable results with regards to early warnings.

Community based disaster management (CBDM) and Preparedness

CBDM and preparedness helps in capacity building of local community to cope with emergency due to disaster. Frequency of disaster increasing day by day and hence it is important to develop and implement a locally appropriate strategy for disaster management. It is therefore necessary that everyone is made aware of prepared to counter disaster, natural or manmade.

Through community approach, it is possible to development their resilience and capabilities. During a disaster. Neighbor and local people are the first who respond the disaster, and they are the most important source to save lives and property in the critical initial moments.

Disaster have wide spread effects and it is not possible to reach at each and every one immediately after a mishap so if community is prepared to respond the govt. machinery will also works more efficiently and losses are contained. In this context the community based Disaster management is emerging as an important aspect all over the world.

Community based disaster preparedness includes involvement of community in disaster management activities, in a way that make them aware to address the root causes of their vulnerabilities by transforming social, economic and political structures that generate inequality and leaves them susceptible to further disasters. (Salajeghel and Pirmondi, 2013)

Community based preparedness is an alternative approach to develop mitigation strategies in consultation with local community using techniques and actions, which they organize themselves and manage with limited technical assistance, these community based programs are more helpful for the development of perception of a community towards the hazards it faces and its ability to protect itself in the future.

Community Based Organizations (CBOs): These are informal organizations emerge as the initiative for the community for some common interest, these are non-government organizations develop to meet certain well defined targets. These are informal organizations also which emerge from the community itself for example formation of a *jungle bachao andolan, Pani Panchayat, Mahila Mandal,* youth clubs etc. There are many other self-help groups which are maintained by these organizations for their own identity and status and enjoy functional autonomy.

The approach promoted under the head of Community-based Disaster Management is to create safety culture in the society and build people's capacity of coping with disaster risks and reducing their vulnerabilities. In this process there is a need of change of focus from emergency response to planned activities that are helpful in the mitigation and prevention of disasters. There is a need for community based disaster management (CBDM) because a well-informed society is far more safer than an obvious one, the need for community involvement in disaster management is also stems out from the inadequacy of approaches in addressing the vulnerable communities, the frequency of disasters is increased in last few decades and disaster related losses are also increased manifold. This approach directly involves vulnerable communities in planning and implementation of mitigation measures for disasters. The approach has gain wide acceptance because communities are considered the best judges of their own vulnerability and are in better position to make decisions for their well-being and can create a more safer and resilient society.

The activities can be decided on the basis of a comprehensive assessment of hazards and an analysis of their specific vulnerabilities and capacities, in these projects and programs of CDBM, resources available locally must be optimally used and then disaster mitigation planning will be sound. This will contribute in strengthening society's security and sustainable development.

CBDM certifies the importance of community as the primary stakeholder in risk reduction and development processes.

Status of CBDM in India

Only few states in India like Gujarat, Orissa and Uttaranchal no other state have initiated CBDM as a state policy to counter disaster. Andhra Pradesh have initiated the Andhra Pradesh cyclone hazard Mitigation projecting, they also involve *Gram Panchayat* in the coastal zone management.

A number of NGOs such as OXFAM-India, DMI Gujarat, Care Andhra Pradesh, west Bengal voluntary health association, Kolkata, VIKAS- Orissa etc. are engaged to mobilize community for participation in the disaster management activities. A number of NGOs are also working for the cause. Mostly these organizations work for various issues but their contribution to preparedness is very meager.

Disaster Mitigation

It is a term used to define all activities to reduce the impact of disasters that can be done in pre-disaster period including preparedness and risk reduction measures. It includes planning as well as implementation phase to manage the disasters both natural and human induced. Generally mitigation is planned after identifying elements at risk.

A mitigation plan includes safety measures for human and animal life, safety of crops and infrastructural facilities and protection of economy from disasters.

The principle aim of disaster mitigation as mentioned in the United Nations Disaster Reduction Organization is (UNDRO) manual titled "Mitigation of Natural Disasters"

(1) Creating awareness for disaster risks at community level.

(2) Promotion local actions through community participation for risk reduction.

(3) Induction of mitigation measures in the national and regional development plans, land use planning and in project design and appraisal in hazard prone areas.

(4) To assist decision makers to understand the nature and extent of various risks faced by communities.

(5) Demonstrate ways and means to reduce those risks within the limits of national, regional or socio-economic conditions through proper decision making and planning.

(6) Introduce effective measures to implement disaster mitigation plans at different level of public administration based on risk assessment and vulnerability analysis.

Types of Mitigation

Mitigation measures which are generally under taken are of two types:

(a) Structural Measures

(b) Non Structural Measures

The structural mitigation consists of engineered structures and non-engineered structures.

The non-structural measures include legal framework of laws and bye laws, incentives, insurance, training, education and public awareness, institution buildings, warning systems etc.

Structural Mitigation:

(1) Engineering and Construction Measures: Through these measures two type of structures are made:

(1) Resistant buildings and

(2) Disaster protection structures such as flood control structures, dikes levees, infiltration dams etc.

Design standards, building codes and performance specifications are important for facilities designed by engineers the engineered designs against the various hazards may include design for vibration, lateral roads load surcharges, wind loads, flood resistance and other safety points. Building codes are most important guideline for engineered structures including large private buildings, infrastructure, transportation networks and industrial facilities.

A code has to fit into an environment prepared to receive it. The engineering mitigation measures also include a high level training for engineers and designers, explanatory manuals to interpret the code requirements. The structural measures that are needed to improve the disaster and non-structural measures involve the education of builders in practical construction methods. For response to cyclone hazard cyclone wind resistance houses are ultimately depend on how roof sheets are nailed down. Examples show that flood control measures which can be adopted by community based organizations prove effective in risk mitigation and simultaneously achieve other development goals, they often use local materials, labour and resources to build on traditional mitigation knowledge. Such measures can play a significant role in disaster mitigation.

(2) Physical planning measures: Effects of localized disasters like floods, landslides etc. can be reduced by avoiding the settlements and construction of important structures in hazardous areas. Most of the urban master plans involve land use zoning try to separate manufacturing structures from the population centres.

Urban planning integrates awareness of natural hazards and disaster risk mitigation into the normal process of planning and development of cities. An important issue is deconstruction of elements at risk like communication lines, hospitals, telephone exchanges, power transformers etc. are important for the proper functioning of a town.

Creation of parks, open lands provides space in the city and reduces urban densities, green belts allow drainage to decrease the risk of floods, and they also provides open space for tents during the emergency period in the event of disaster.

The design of pipelines and cables needs a complete planning to reduce risk of failure, long supply lines are not available as they are at risks. if they are cut at any point.

By changing of land use of a vulnerable building the urban planner can reduce risks for example a departmental store in a weak building can be moved to a stronger one weak building can be used for the functions like storage.

The private infrastructure and buildings are difficult to control than government buildings through urban planning therefore they are at great risk of incurring damages caused by disasters.

Low-lying areas and other lands prone to various hazards are generally reside by poor people of low income group. The poor people living in these marginal lands are the most vulnerable social group of the society.

Economic Measures:

(3) Economic development of a region is the most important for disaster mitigation. A diversified economy in which various activities have share and their benefits are also shared is perfectly protected against a disaster.

A strong and rich economy has more capital to spend on infrastructure and larger capital to cope with future losses. The people with least resources proportionally, loose more in disaster. The weaker group of society have few economic resources, only their houses and animals if they lose it than there is no mean to recover it.

They don't have facilities such as insurance and access to credit and therefore quickly become destitute.

Economic development is always foremost objective of any regional plan and government department, the mitigation process will boost up as it is a part of it. Various angles of economic planning are directly relevant to reduce disaster risk. An economy based on single activity is always more vulnerable than a diversified economy with many activities.

Economic incentives, grants, loans, taxes and fines are influence people's decisions to reduce disaster risks. In the developed countries insurance is one of the major economic protection measures. Commercial insurance is expensive and it depends upon the calculation of risk.

Many countries through general taxation has already created a disaster reconstruction fund in their territory.

Disaster insurance is high risk finance and multinational insurance companies only has the resources to pay for the losses.

4. Institutional Measures:

Disaster mitigation requires certain organizational and procedural measures. The objectives and policies formulated for mitigation processes have to be sustained over a number of years.

The institutionalization of disaster mitigation requires a consensus that efforts to reduce disaster risk is of utmost importance. Political will is a necessary aspect of institutionalization of disaster mitigation.

Training of engineers, planners, economists, social scientists and other managers is compulsory to include hazards and risk reduction in their area of competence.

Internal exposure to these is very important as it increases knowledge by transfer of technology in disaster mitigation which is a crucial part of capacity building in the affected country.

Information is very crucial element in disaster mitigation, the countries with good information systems are quite resourceful in this respect, where as those which are hazard prone but lacking in basic meteorological and geological observations to monitor hazards.

Administrative and organizational powers include the checking procedures and planning powers to realize mitigation plans, consultation methods and involvement of the community in disaster mitigation plans and implementation of mitigation activities.

5. Societal Measures:

Involvement of society in the mitigation activities has its implications as many time as it observed that a well-informed society is less vulnerable to disasters than an uninformed one, sometime people don't know the threats and therefore they are not recognized, the prevention step that people can take to protect themselves and their community as much as possible.

For awareness generation various steps can be adopt, from a short term to long-term basis. High profile campaigns using media reports, literature and posters and low profile campaigns that are carried out through general education. Everyone living in a region should know about the hazards as a fact of life. Information of hazards should be part of everyday information sources with mentions in stories, TV serials and programs, news channels, newspapers and magazines. In this way people should be aware of pros and cons of disasters and the safety ways to protect themselves and not terrified of the possibility of hazard occurrence, their choice of houses, the placement of the bookcase, stove and construction codes, all affect their own safety.

The primary aim is to reduce losses in the event of disaster by minimizing life and property loss.

Secondary aim is to reduce economic losses and damage, it can be achieved by various measures such as spatial planning, engineering measures, and management societal inputs will be needed to bring about effective mitigation. It can be defined as the tendency of a thing to be damaged by a hazard. People's life and property are at risk directly from the destructive effects of the hazards. The resources are at risk because of destruction of buildings, crops, livestock and equipment.

During disaster phase

Response

It is to ensure that the needs and demands of victims are met to alleviate and minimize suffering. This refers to first stage of response to emergency, which includes setting up control rooms, issuing warnings, evacuation of people and moving them to safer areas and providing medical aid to people in need. In this phase most importantly immediate relief, rescue and the debris clearance is done.

Major sudden onset natural disasters such as earthquakes, tsunamis, cyclones, floods typically damage infrastructure and cause mass displacement of people and also injuries and deaths. People trap in debris. An immediate response is required

This phase includes efforts for rescuing people and providing shelter to the victims, shelters are temporary arrangements for people who are evacuated from their homes after occurrence of disaster. Removal of debris and recovering people who are injured or deceased also is a very major task. This phase has three important elements – Recovery, Rehabilitation and Reconstruction.

Recovery includes all the acts done to recover the people and removal of debris .An immediate response is required to rescue people many commence with search and rescue of victims and safeguard them at safer places or temporary shelters. This assistance is provided by national and international organizations, in the post disaster period, the foremost relief operation which are search and rescue. The difference between life and death can be a matter of hours. Initial lifesaving search and rescue must be very fast, possess specialized skills some time needs heavy equipment. The operation is highly challenged as responders work in places covered with debris and damaged structures or water logged areas.

In the natural disasters, the goal of a search and rescue operation is to evacuate largest number of people in the shortest time.

Search & Recue

Search and rescue are complex tasks as peoples hide due to disaster in various locations the techniques are focus on the places where people are trapped or hide intentionally. The areas of entrapment inside damaged structures are called voids, under which victims feel protected such as tables, cupboards, desks, pillows etc. Search teams make different strategy to search the victims one such method is called systematic triangulation in this task three people approach an entrapment zone from three directions.

Second method is a right and left search pattern in which one team searches the left side and the other team on the right side of a building.

Important steps for search

Carefully examine the damaged structures, mark the landmarks and draw a rough sketch this layout will help in deciding the rescue path, rescue area & method to be adopted for search operations. After making a search plan we can easily move to our task. To search people and Provide the first aid to victims.

1. **Equipment**: Many advance equipment are now available for search and rescue operations one such equipment is Life guard made by USA designed by ex NASA scientists, used by many countries of the world including India(NDRF) this equipment looks like a gun, a laser light is come out when we use it this equipment is sensed with living human heart if there is any living person inside the debris it can pointed out the location of that person from outside, hence its very useful in searches in areas where people are affected by earthquake, landslide, Avalanche, collapse of a building etc. As it save time and energy of the rescue team and vice versa save the life of victims.
2. **Victim location search camera**: it's a probe mounted camera to check for the live or dead victims inside nay collapsed structure, as with its use one can see the situation inside before entering and accordingly advice the team to take further action.
3. **Air lifting bags:** These are high pressure bags which can sustain high weights during any earthquake or collapsed

structure and rescue team can use it as a jack to create space and hold it during any rescue operation.

4. **Under water Rescue equipment:** Diving equipment are also very useful for under water rescue during flood operation or any other incident.
5. **The personal protective equipment s (PPEs)** like; Hamlets, Masks, Gloves, SCBA Sets, Goggles, Protective Suits etc.
6. **A complete medical aid kit.**
7. **Communication System.**

Trapped disasters victims are rescued and given medical aid if necessary, before removing victims it is essential to remove debris so that victims may be able to take out safely or may be removed by drags or carriers etc.

The people are also searched by sniff dogs, helicopters boat crafts etc.

Generally a sector is demarcated for each search and rescue team in the case of predictable slow onset disasters they are geared up and place in a stand by position to get into action as and when required.

Trained rescuers are required for the process of evacuation which involves mainly the relocation of the disaster affected population from areas of risk to a safer destination for successful evacuation to be happened an early and accumulated system is required.

A comprehensive policy is required for search and rescue by their public education can create awareness towards evacuation.

1. **Search of Building on Fire:** The leader shall ascertain from neighbors, if there is anyone still trapped in the effected building. If possible rescue team should work in pairs. Start searching from the top of the building, so that the searcher may be nearer the fresh air at the end of search operation. Search swiftly but thoroughly. Do not avoid any possible hiding place.
2. **Search of Smoke Filled Room:** Do not open the door of the smoke filled room suddenly may be the heated gas & smoke inside may overwhelm the fire fighter personnel, open the

door slowly in a crouching position, this will allow the hot smoke and gases to pass over the head. If the door of the room opens outwards, place one foot against the bottom of the door & open slowly .This will protect from injury as high pressure develops result force on the door due to the expansion of the heated gases inside the room. Enter the room in a crawling position as there is less smoke and less hot air near the floor. Keep a piece of cloth on the nose. Keep yourself near the wall while moving. Make a complete circle in the room. Find victims under the beds, inside cupboards, wardrobes and other likely places where one can hide. Cross the room diagonally to make sure that no one is lying in the middle.

3. **Victim Management**. Continue search till the last victim is rescued.
4. **Precautions during Search**. Keep your comments on positive note. Always assume someone is listening to you. The victim is in the worst possible position & fighting to stay alive. You can enhance his chances of survival by being positive about the possibility of finding & saving him. You may be first person the victim is able to communicate with; therefore it is important to project a sense of confidence & hope.
5. **Steps for Initial Contact with the Victim.** Try to talk in a language familiar to victim, introduce yourself as a rescuer, showing confidence and be calm in your voice and choice of words.
6. **Search Strategy**. This involves developing a process for detecting and locating the greatest number of victims in the shortest amount of time period.

   This is composed primarily of physical and canine search operations.

   Search Plan: a search plan will ensure that all the search group efforts will be coordinated in a systematic manner by developing a search plan based on the results of area

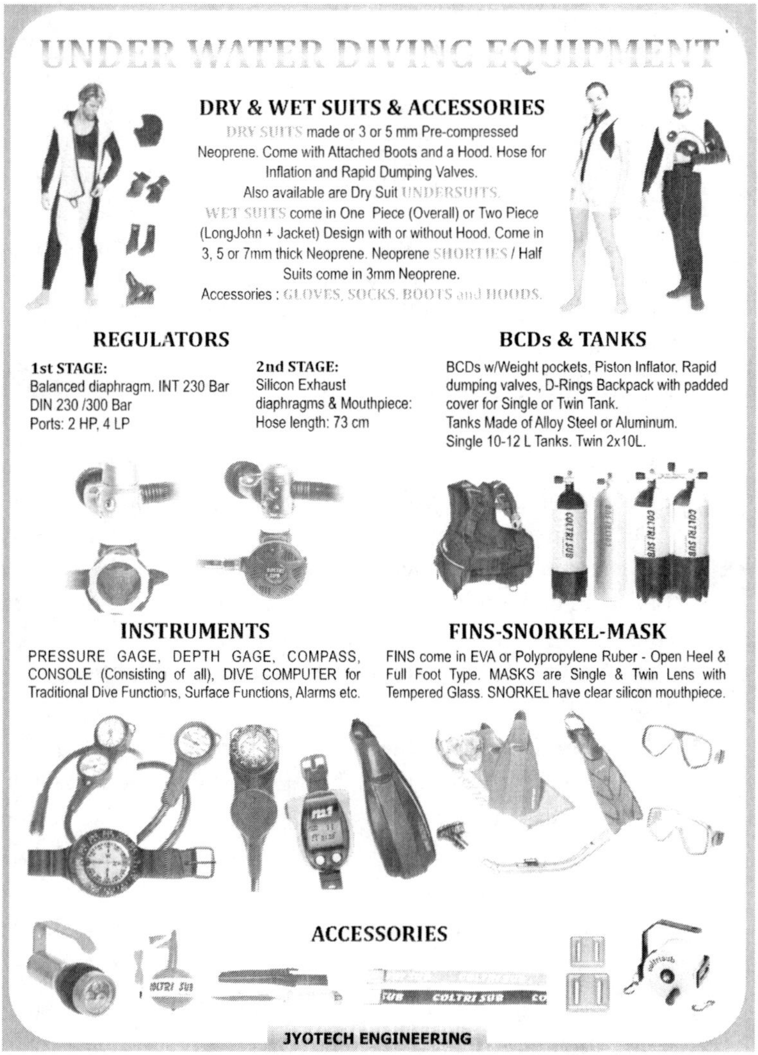
UNDER WATER DIVING EQUIPMENT
DRY & WET SUITS & ACCESSORIES
DRY SUITS made or 3 or 5 mm Pre-compressed Neoprene. Come with Attached Boots and a Hood. Hose for Inflation and Rapid Dumping Valves.
Also available are Dry Suit UNDERSUITS.
WET SUITS come in One Piece (Overall) or Two Piece (LongJohn + Jacket) Design with or without Hood. Come in 3, 5 or 7mm thick Neoprene. Neoprene SHORTIES / Half Suits come in 3mm Neoprene.
Accessories : GLOVES, SOCKS, BOOTS and HOODS.
REGULATORS
1st STAGE:
Balanced diaphragm. INT 230 Bar
DIN 230 /300 Bar
Ports: 2 HP, 4 LP
2nd STAGE:
Silicon Exhaust diaphragms & Mouthpiece:
Hose length: 73 cm
BCDs & TANKS
BCDs w/Weight pockets, Piston Inflator, Rapid dumping valves, D-Rings Backpack with padded cover for Single or Twin Tank.
Tanks Made of Alloy Steel or Aluminum.
Single 10-12 L Tanks. Twin 2x10L.
COLTRI SUB
INSTRUMENTS
PRESSURE GAGE, DEPTH GAGE, COMPASS, CONSOLE (Consisting of all), DIVE COMPUTER for Traditional Dive Functions, Surface Functions, Alarms etc.
FINS-SNORKEL-MASK
FINS come in EVA or Polypropylene Ruber - Open Heel & Full Foot Type. MASKS are Single & Twin Lens with Tempered Glass. SNORKEL have clear silicon mouthpiece.
ACCESSORIES
COLTRI SUB
JYOTECH ENGINEERING

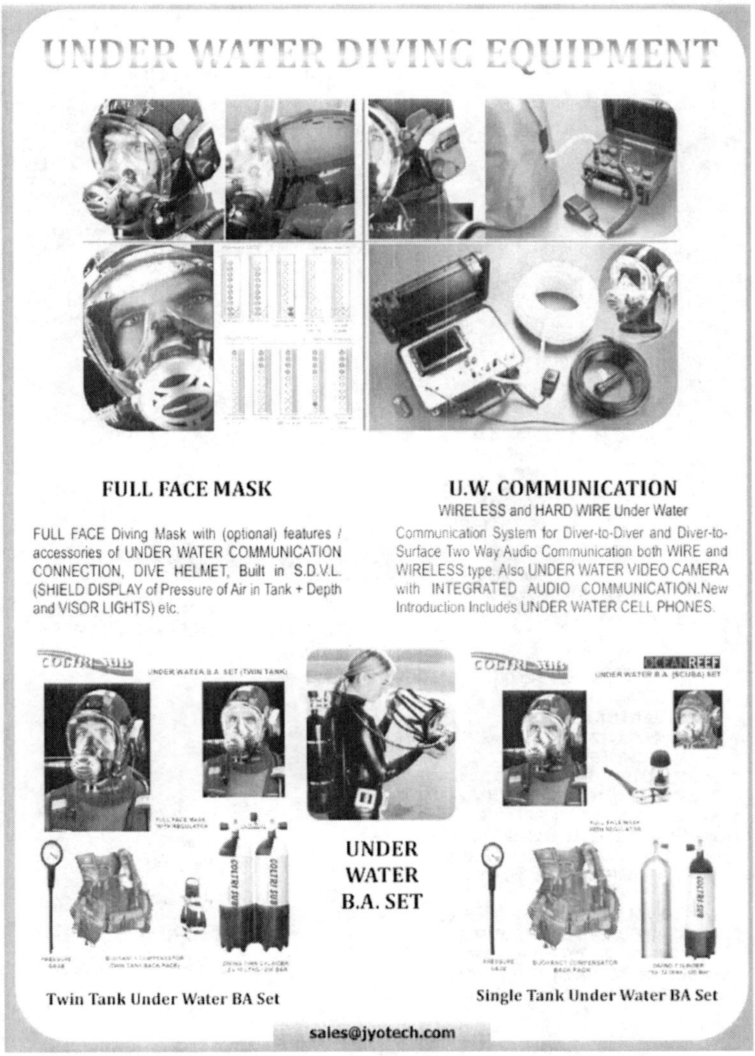
UNDER WATER DIVING EQUIPMENT
FULL FACE MASK
FULL FACE Diving Mask with (optional) features / accessories of UNDER WATER COMMUNICATION CONNECTION, DIVE HELMET, Built in S.D.V.L. (SHIELD DISPLAY of Pressure of Air in Tank + Depth and VISOR LIGHTS) etc.
U.W. COMMUNICATION
WIRELESS and HARD WIRE Under Water Communication System for Diver-to-Diver and Diver-to-Surface Two Way Audio Communication both WIRE and WIRELESS type. Also UNDER WATER VIDEO CAMERA with INTEGRATED AUDIO COMMUNICATION.New Introduction Includes UNDER WATER CELL PHONES.
UNDER WATER B.A. SET
Twin Tank Under Water BA Set
Single Tank Under Water BA Set
sales@jyotech.com

SEARCH & RESCUE EQUIPMENT
LIFEGUARD™ HUMAN LIFE DETECTOR
DISASTER SEARCH
FIRST SEARCH TOOL TO LOCATE SURVIVORS TRAPPED UNDER DEBRIS IN CASE OF EARTHQUAKE - BUILDING OR STRUCTURAL COLLAPSE, LAND SLIDE, FLOODS - PEOPLE TRAPPED UNDER WATER - FROM OUTSIDE
MINE RESCUE
SEARCH FOR MINERS TRAPPED IN FLOODED OR COLLAPSED MINE FROM SURFACE.
SECURITY SEARCH / BORDER PATROLLING
LOCATE PEOPLE HIDING BEHIND BARRIERS, IN BUSHES & BUILDINGS FROM A DISTANCE / COMBING ACTIVITIES
RANGE : UPTO 500m
DETECTS ONLY LIVING HUMAN BEINGS (No Animals)
NO POWER SUPPLY REQUIRED
PROEYE 951.S / 951.S.IN SERIES
VICTIM LOCATION SEARCH CAMERAS
MAIN FEATURES
COLOR VIDEO AND 2-WAY AUDIO SEARCH FEATURES
TELESCOPIC PROBE WITH PISTOL STYLE CONTROL GRIP
REMOTE CONTROLLED TWO-WAY ARTICULATING CAMERA
CAMERA HEAD WATERPROOF TO 50M
5.4" FLAT LCD DISPLAY ON THE PROBE
VARIABLE INTENSITY LED ILLUMINATION
RECHARGEABLE BATTERY PACK WITH CHARGER
PACKED IN A WATER AND DUST TIGHT CASE
OPTIONAL ACCESSORIES
REMOTE WIRELESS / HARD WIRE MONITOR
BREACHING SYSTEM TO DRILL HOLES IN RCC / CONCRETE
CABLE FOR UNDER WATER / DEEP SEARCH (20 / 30 / 50m)
INFRA-RED THERMAL SENSOR TO CHECK BODY TEMPERATURE
System 751DXS
BREACHING SYSTEM
JYOTECH ENGINEERING

SEARCH & RESCUE EQUIPMENT
AIR LIFTING BAGS
LOW PR. BAGS (0.5 BAR)
ROUND BAGS - UPTO 09 TONS CAPACITY
UPTO 2.1 METER LIFTING HEIGHT
MEDIUM PR. BAGS (1.0 BAR)
ROUND BAGS - UPTO 12 TONS CAPACITY
UPTO 1 METER LIFTING HEIGHT
HIGH PR. BAGS (8.0 BAR)
SQUARE / RECTANGULAR BAGS
KEVLAR® CHORD RE-ENFORCED
UPTO 65 TONS CAPACITY
UPTO 0.51 METER LIFTING HEIGHT
FLAT FORM BAGS (8.0 BAR)
COLLECTION OF 3~4 SQUARE BAGS
KEVLAR® CHORD RE-ENFORCED
UPTO 65 TONS CAPACITY
UPTO 0.19 METER LIFTING HEIGHT
AIRCRAFT RECOVERY BAGS
COMBINATION OF 3 & 1 CHAMBER BAGS
CAPACITY TO 45T. OZONE / HC RESISTANT
SERIES OF BAGS FOR COMPLETE AIRCRAFT RECOVERY
LIFE EXPECTANCY OF 15 TO 20 YEARS
ACCESSORIES
CYLINDERS, PRESSURE REDUCERS
SINGLE / DUAL / TRIPPLE CONTROLLERS
FOOT PUMP, HOSES & SHUT-OFF VALVES
INFLATABLE STRETCHER
2.4 x 0.7 x 0.37m m (0.3 BAR PRESSURE)
INFLATABLE SHELTERS
12 TO 52 SQ.M. AREA. OTHER AVAILABLE
SET UP TIME LESS THAN 15 MINUTES
REMAINS FULLY INFLATED FOR DAYS
MADE OF FIRE RESISTANT MATERIALS
USES - DECON SHELTERS, HAZMAT STORAGE
INCIDENT COMMAND CENTER, STORAGE OF
FIRST AID CENTER, MOBILE HOSPITAL ETC.
INFLATABLE WALKWAYS
5 & 10 M LENGTH. CUSTOM SIZE POSSIBLE.
OTHER PRODUCTS
WATER WALLS, DE-CONTAMINATION UNITS.
LEAK SEALING PADS, PIPE SEALING PLUGS etc.)
sales@jyotech.com

assessment, hazard assessment, etc. The search plan should include search strategies and tactics, maps and drawings of the concerned area, safety and security considerations.

International search and rescue advisory group: As a cooperative effort by the United Nations the group was formed in 1991. The participants of this group have developed a common understanding of the functions and operations of search and rescue teams which have resulted in the development of the search and rescue system.

Relief

Disaster relief is the post disaster response to a catastrophic situation, by this process aid is provided to victims and communities who have suffered from some form of loss and injury. It involves dealing with the needy and avoiding risks, supporting, and rebuilding society when natural or human induced emergencies occur. Effective relief management relies on the integration of emergency plans at all levels of government and non-government organizations. Commonly this is the responsibility of government with the institutions for civil defense or within the conventional structure of the emergency services. International organizations, such as the Red Cross, provide aid to disaster victims throughout the world.

Relief is a quick administrative response from centre and state levels. Immediate deployment of defense and national disaster response force, spontaneous public response with overwhelming generosity in relief assistance and distribution. Quick delivery of food items, drinking water medicines etc. we need a holistic approach, water, health, food, shelter, livelihood are the essentials which are required by the victims as disaster disrupt all these facilities.

Disaster relief require various steps to follow and measures to be taken. An integrated framework for disaster relief will necessitate proper planning and its enforcement in practical terms. There **are four phases of relief work:**

The first one begins when a disaster strikes, through the all phases efforts of future prevention and risk reduction are continue.

Each phase is characterized by specific activities and requires varying amounts of support in terms of both time and money, **Phase 1** begins immediately after the disaster strike focusing on restoring order to the area.

Phase 2 quickly follows as residents work toward stabilization and a return to daily activities to create normalcy.

Phases 3 and 4 are characterized by community rebuilding and rehabilitation and ongoing awareness generation and preparedness education in order to mitigate future disasters by building better and safer livelihoods, creating a safety culture by informing the community about the threats.

Preparing for disaster relief include the following:

1. Shelter
2. Resettlement, relocation camps
3. Food
4. Water supply
5. Power supply
6. Health and Sanitation
7. Public information
8. Security
9. Safety and welfare
10. Debris removal
11. Disposal of the dead bodies
12. Livestock relief measures

Shelter is most important necessity in post disaster period especially when physical damage is caused such as after an earthquake and cyclone. In case of floods also evacuation takes place and rescued people needs temporary shelters especially built for the purpose of rehabilitation, the shelter or camps being established should be disaster specific site specific and culture specific.

Livestock related relief measures need to target to reduce the loss of animal life, preventing epidemics, arranging fodder for animals and active animal husbandry in post disaster period with proper stock of medicines, related facilities and availability of doctors.

Disposal of dead bodies of both human and animal is a major relief operation. This assignment needs urgency as decomposing dead bodies lead to epidemics and disease.

Debris removal is also an elaborated work as many type of buildings, flyovers, bridges, uprooted trees, electric telephone booths, damaged vehicles and goods, accumulated solid waste etc., are there to removed, the clearance work is very troublesome and time consuming which require electronic devices, flashlights and various types of machines.

Relief Management

In order to ensure timely delivery of that relief material to the disaster struck area, following steps may be adopted:

Designate person or groups within the supply section as the focal point for the given disaster.

There should be emergency arrangement to minimizing disruption of on-going supply functions while meeting emergency needs.

Coordination among the concerned departments and ministries with other partners and counter parts to streamline supply procurement to avoid duplication.

Special task forces should be included to finalize and establish correct products, specifications, quantities, qualities, distribution and storage systems.

Establish an active communication mechanism that provides up-to-date about the supplies, all the stakeholders must have access to the latest information.

Monitoring of all activities and feedbacks should be reviewed for future.

First Aid Management

First aid is a medical aid which provide to a patient before the arrival of a doctor. During disaster when emergency occur many people injured and some of them dies if not attended immediately, thus medical aid in post disaster period is mandatory to save human lives and prevent epidemics and disease to spread. The physical disasters earthquake and landslides and collapse of building structures can cause serious head and bone injuries in that case the disaster managers must know how to stop the blood and take the patient immediately to the hospitals and dispensaries.

Disasters create trauma and emotional problems some people are caught by cardiac arrest and fell unconscious in that case the first aid team can do the following treatments;

Response

First check the response of the victim by asking name and squeeze shoulders, make patient comfortable and monitor breathing if patient is not responding transfer him to the nearby hospital.

Breathing

Check breathing if patient is not breathing normally start cpr, if patient start normal breathing monitor breathing.

CPR: is an emergency lifesaving treatment given to a patient when his or her heart stops beating. Immediate CPR can increase the chances of survival many times after cardiac arrest.

CPR helps in keeping the blood flow active extends the chances of resuscitation if given by a trained person immediately after a heart attack.

If patient is not conscious he is not breathing then start CPR compress the chest and start pumping by hands over the chest continue CPR until help arrives or patient recover.

Ready Medicine kit for

Infection control, gastro, pain relief, allergic reactions, asthma attacks, bites and stings, severe blooding, fractures and dislocations, spinal injuries, head injuries, heart attack, poisoning, burns, choking, shocks etc.

First aid is the first help to the patients it can prevent life of people. All the people engaged in disaster relief should be trained to give first aid to the victims.

Post disaster Rehabilitation & Reconstruction

Rehabilitation refers to the actions taken in the post disaster phase to restore functioning and are self-help efforts to repair the damages and community facilities, revive economic activities and provide help to the survivors and bring them back in the mainstream and in the pre-disaster pattern of life. Rehabilitation includes temporary facilities of housing and utilities as interim measures for long term recovery.

Reconstruction involves return of victims permanently to their houses which are repaired or reconstructed after the disaster, replacement of buildings and public utilities is done so that sustain long term development goals.

Rehabilitation include retrofitting and rebuilding of the damaged housed and other structures, these are important issues which are needed to be addressed in post-disaster period. People are in need of social, economic and other rehabilitations.

The social and psychological rehabilitation of the affected people are a community function, as many people such as elderly without a family, orphans and single parents with young children are most vulnerable due to lack of support. These people need special support to survive after the disaster.

These people need special support to survive after the disasters.

Strengthening of healthcare facilities and infrastructure is very important at the time of emergency period during disaster or during the outbreak of an epidemic in post-disaster phase.

For this co-ordination of medical support muse be essential in vital locations, networking among various health departments, medial education centres and hospitals. Control centres can be used for collection and storage of medicines from national and international relief agencies.

Reconstruction of education institutes and schools in the affected areas is imminent task, counselling of teachers to and students is also required to start education process. Necessary materials like books, writing material, transport facilities are also made available.

Rehabilitation of most vulnerable group – women, children and elderly people in post-disaster period is a matter of great concern, in every disaster this group suffers more as compared to others. Activities included are development of physical infrastructure for women and children such as Anganwadi, community centres and female children homes. Generation of economic opportunities for women from vocational training and self-help groups.

Resettlement and rehabilitation of women and children should mostly be done in a safe and familiar environment. Long term rehabilitation requires good infrastructural facilities, finance and adequate resources.

Economic rehabilitation focuses on economic recovery and attainment of same growth level as it was in pre-disaster period, so as to bring things back to normal.

Agricultural rehabilitation is done to overcome losses necessarily in rural areas, it includes short term measures such as providing seeds to sow for upcoming crop season to farmers or help in harvesting of the crops, provisions for availability and distribution of fertilizers, pesticides, etc.

Long term measures are also there such as provision for large number of agricultural equipment and implements in addition to other agricultural inputs, facilities of irrigation of wells and pump sets. Can be ptovided There must be some scientific studies carried out to assess the actual situation of required system and their urgent rehabilitation.

Improvement of shelters, infrastructure in relocated areas, roads and bridges, public buildings, repair and retrofitting of historical monuments etc.

Reconstruction and rehabilitation are a tool for social development, economic development and environmental restoration such as afforestation.

Reconstruction refers to the full restoration of all services, replacements of damaged physical structures and the revitalization of economy. Reconstruction is integrated into developmental plans.

Reconstruction always take into account future disaster risks reduction measures. The structures and services already damaged may not be restored in their previous forms.

The physical reconstruction is essential for return to normalcy, it is a quantifiable visible achievement of the authorities and donors.

A comprehensive reconstruction plan should take into account both physical and non-physical needs of the communities.

Ideally, reconstruction aim to build to a better standard.

Reconstruction and Development

Post disaster reconstruction can influence development programs both positively and negatively. Delays in reconstruction will also divert public and private investments and resources away from development activities, reconstruction programmes need to be planned.

Through developmental programs we can change the vulnerable conditions for the high risk population.

Role of Various Agencies in Reconstruction

Comprehensive rehabilitation measures are usually taken up in case of high magnitude disasters with the assistance of International agencies by government of the country where disaster occur can take up rehabilitation program, for restoration of shelter and livelihood required in the affected area. The resources are move through the concerned ministry as per the needs to counter disasters appropriately.

Governmental Agencies: Central government provide physical and financial resources, revenue and Relief departments are responsible for natural disaster. At the state level important decisions are taken by chief ministers collectors are supervise relief measures.

Non-Governmental Agencies: Government alone cannot achieve the target without the help of local organizations. NGOs tend to work with privileged. NGOs activities play an important part of international assistance and relief work. NGOs work with the needy as well as the marginalized groups in the society. International NGOs like the Red Cross Society, CRY, UNICEF etc. Work for disaster victims in the specific areas of field operations, and financial, as well as technical support.

The NGOs are growing worldwide rapidly mainly in the past two decades, largely due to the helping hands for government as well as to overcome the shortage of resources and manpower specially at the time of emergencies, therefore it is now believed that NGOs are an effective actors in disaster management activities.

Some important NGOs are:

Cooperation for Assistance and Relief Everywhere (CARE India): It is work to recover human sufferings and support systems in disaster situations. They work closely with national and state governments.

Oxfam India Trust: It works in collaboration with other organizations, and provide funds for the rehabilitation programs in disaster affected areas.

Indian Red Cross Society: In the event of a disaster, the services of the Red Cross become complementary to that of the government. It provides immediate relief in case of disaster. It takes care of medicines and first aid needed for the disaster affected people.

International Agencies

The important agencies are: United Nations Development Programme (UNDP) United nations Economic and social Cooperation (UNESCO), United Nations Children's Fund (UNICEF), World Health Organization (WHO), United Nations Fund for Population(UNFPA), International Labour Organization (ILO) World food programme. (WFP).

International Federation of the Red Cross (IFRC) and Red Crescent Societies:

It is the world's largest humanitarian organization. Its main mission, broadly speaking, is to improve the situation of the world's most vulnerable people.

USAID: United States Aid Programme doesn't directly involve itself in disaster relief, it is more concerned with monitoring and reporting on disaster preparedness and response activities. There may, however be a provision through partner NGOs for short term food aid to disaster victims.

World Bank: Streamline disaster prevention and mitigation initiatives connected with all the activities pertaining to disaster management. It has been doing a lot of work in the area of disaster mitigation.

UNICEF: It donates funds to the prime Minister's National Relief Fund. It regularly meets with the government of the affected country in order to facilitate a more coordinated disaster response mechanism.

UNDP: According to agreement UNDP does not allow to participate in relief operations unless specially requested, it does receive all information on the nature and extent of disaster damage and loss of life. Its broad aim is to design responsive structures as well as implement prevention and mitigation strategies.

Problems: There are many problems that are faced by the planners in the planning for rehabilitation and reconstruction. The actors involved in the process are central government, local authorities and departments, NGOs and community. The losses should be strictly evaluated from damage of agriculture or business premises may be evaluated by the economists, inspection of damaged buildings by the engineers. Detailed damage assessment can help in understanding the cause of damage and the resources of risks and vulnerability.

Chapter - 4

Risk

Risk is the probability of occurrence of an event. Risks of varying types and magnitudes exist everywhere. However, some risks are far greater and be disastrous if not identified and properly managed in terms of reducing the probability of risk.

The initial step in disaster preparedness is to identify the potential risks, assessment of risk, evaluation of probability of occurrence of risks and estimation of potential damages. Potential risks vary from operation to operation and location to location.

Identification of risk is a process by which we determine risks that can cause danger to life and property.

Risk identification is the first step of disaster management process. The main objective of risk identification is to identify those areas where disaster can occur.

Identification of risks include listing of natural and man-made hazards that could potentially have an adverse impact.

Risk Assessment

Risk management according Coburn, Spence and Pomonis(1994) has two components: Risk assessment and Risk Evaluation.

Risk assessment is a scientific process of data collection and maintenance over time and its analysis to manage disaster situations and planning of preparedness according to the requirements. First step for risk assessment is to identify all the hazard and vulnerabilities of the region. This type of assessment together provide a picture of the risk. Risk assessment includes a close examination of all the relevant elements that are vital for the understanding of existing hazards and their impact on the environmental conditions. About the hazards we get knowledge from the physical sciences like meteorology, hydrology, geomorphology, geology, seismology and volcanology. The factors responsible for vulnerability in a region must be inquire properly and preventive measures can be taken accordingly. Both natural and anthropogenic factors are included in it, these factors are come out due to backwardness of a community in society or lack of resources to sustain life, particularly during disaster. Hazard assessment, also known as hazard evaluation is the process of estimating of risk for defined areas within specified period of time.

Risk assessment is generally carried out in two ways, the quantitative assessment and the qualitative risk assessment. Quantitative or statistical analysis is done to arrive at a peculiar statistical model that relates risk created by a natural catastrophe to socio-economic parameters.

Quantitative risk assessment requires calculations of two components –

Risk (r) = magnitude of potential loss (l) and the probability (p) that the loss will occur.

Risk assessment includes two crucial things – potential loss and probability of occurrence can be very difficult to measure. The chances of errors are very high in measuring these two components.

$r_l = l_l p\ (l_i)$

r_l = total risk

l_l – potential losses

p = Probabilities

Steps of risk assessment:

(1) Identify the hazard that may be an area and responsible for potential adverse consequences.

Techniques of Risk Assessment

Following methods can be used for risk assessment

> An analysis can be carried on to identify threats and hazards, this is a brainstorming activity and carried out by people who have knowledge about the areas, sites and process that may be exposed to hazardous events and situations.

This type of analysis requires a check list to identify hazards. The value of this type of analysis depends upon the quality of check list and the experience of its users.

Check list ensures that all the relevant questions are asked and discussed so as to encourage a creative approach to risk assessment.

Risk assessment steps:

Comprehensive risk assessment

(1) Identify the range of hazards.

(2) Impact of hazard on infrastructural facilities.

(3) Determine the potential impact of hazards

- o Estimating the relative severity of each hazard.
- o Estimation of relative frequency of each hazard.
- o Estimation of vulnerability of surroundings and environment from hazards.

(4) Categorization of each hazard on the basis of severity, frequency of its occurrence and degree of vulnerability.

(5) Develop strategies to deal with the adversities of hazards, to prevent them and to prepare for them and to provide an efficient response followed by recovery from the hazard.

(1) Fault tree analysis In this analysis a risk criterion is decided to follow two main approaches – one is to identify the risk which is permissible to expose an individual to while the other one is to determine the maximum amount of finance and infrastructure facilities required to be spent in order to avoid fatalities (Kletz 1986, Lee 1980a).

A risk analysis is done to demonstrate equipment safety. The situation is described by a simple model. Most of the treatments deal with risks primarily the risk of death. Data on fatalities are recorded. Fatal accident rate is calculated it is defined as the number of deaths per 10 exposed, it means number of deaths over a working life time along a group of thousands of workers. This is also called fatal accident rate.

The risk R to a person can be computed as:

Where, X is the number of deaths in a particular type of potential accident

F is the frequency of such an accident

n is then number of type of potential accidents

N is total number of people at risk (Khan and Abbasi 1985, Lees 1980)

It is a deductive method which is used for quantitative assessments, it is an initial study of the system under discussion. This analysis is carried on to search the shortcomings of as system which can cause a hazardous event, the results are shown graphically with the help of a fault tree. This tree depicts quantitative outcome of the analysis.

This analysis generally has two types of gates – OR and the AND gate. In the case of OR any entry is capable of producing the output event while output of AND gate occurs only if all its entry events are normally included.

In order to carry out a fault tree analysis the following steps are required:

- Complete information about the system.
- Definition of the hazardous events, material information, checklists and past evidences.
- Development of the fault tree.
- Obtaining probabilities for the occurrence if disaster.
- Evaluation of fault tree.

(2) Maximum Credibility Accident Analysis (MCAA)

Risk assessment is complete when the consequences of a possible accident are known. The result of the analysis are used to determine the protective measure to control hazard. An accident consequence should contain the following–

(1) Description of the event.

(2) An estimate of the quantity of material released.

(3) A calculation of the dispersion of the material released.

(4) Estimation of effects – In order to identify the undesirable events, it is advisable to create an artificial site of possible events. Then mathematical modeling of chemical discharge effects based known physical phenomenon is done.

3. HAZOP

The results of a Preliminary Hazard Analysis(PHA) show that a more detailed study is required which presents a significant hazard to operators or to the environment. Fir this a HAZOP study is required to identify systematically the possible ways. The objectives of HAZOP study are:

(I) To identify the areas which may possess a significant hazard potential.

(II) To study features of the design which influence the probability of the occurrence of disaster.

(III) To ensure that a systematic study is made of the areas of significant hazard potential.

(IV) To provide a mechanism for feedback to the client of the study.

4. Hazard Analysis (HAZAN)

Hazard analysis is a method utilized for assessing of various hazards. It recognizes four situations with respect to probability of occurrence of an event which are as listed below –

1 – Low probability of an event

2 – The probability of the event can be made low by the application of a standard.

3 – The probability of the event can be made low by the use of measures which can be assess by risk analysis.

4 – T1980he probability of the event is assessed quantitatively and is reduced to conform the risk criteria.

For simple cases, one can use fatality analysis (FA) while for complex cases fault tree analysis is recommended. For a successful program of disaster risk reduction it is important to understand that a society is vulnerable to which hazards that it confronts, that is its own perception of risk so that decision can also involve in it. Decisions are made and actions are taken according to the way that risks are perceived. Different societies have different Perception of risk.

Perception of risk related to psychological parameters rather statistics, the mental process of evaluating risk of different types of information tends to differ significantly between individuals and groups.

Perception of risk is considerably influenced by availability of information. Some societies without access to information on hazards seems to have poor perceptions of the risk of

potential hazards that might strike them. Likewise, rural communities with little exposure to media and formal education may have less perception of risk, their perception of risk is based on personal experience, verbal folklore and on local and recent events.

Risk Reduction

Past experiences with natural disasters asserts an urgent need of prevention and mitigation from mobilizing resources to reduce the risk before the disaster occurs. In emergency management there is distinct difference between the developing and developed world. Developing countries give emphasis of preparedness and response making sure that they have enough resources to respond to emergencies and to quickly tackle when such situations arise. Developed countries on the other hand give importance to reducing risks and mitigating the effects of disasters.

Disaster reduction programs can reduce the losses of life and property by long tern mitigation measures. There are several instances for example after installation of flood control dams and improvement in drainage systems, the Rio flood reconstructions and prevention estimated 6.5 benefit cost ratio for seven sub-basins of Iguacu and Sarapui rivers. Comprehensive risk management can be integrated into development. In Turkey also the international lenders and donors worked with government to develop a new disaster management framework in the aftermath of the 1999 earthquake.

Reconstruction of settlements in the flood prone or volcanic areas. In these programs inhabitants of low income urban areas improve their houses themselves or with the community's help. Other construction activities include construction of drainage lines and reducing the risk of flooding and mud slides environmental conservation and reforestation can also reduce vulnerabilities of natural disasters. For drought and flood prone areas, community food banks can help.

Technologies provided alternative livelihoods for those living in degraded landscapes (Slater and Levin1981).

However, technology not always helps in reducing vulnerability. Irrigation can cause conversion of productive lands into waste lands through water logging and leaching of soils. In drought year, the vulnerability further increases (Heath cole 1983).

People's Participation in Risk Assessment

People's participation is so important in countries like India because process of policy formation in these countries takes place in the ecological context.

Participation also has subjective value in that a feeling of self – reliance is inculcated in subject communities, community participation encourages people of all classes to be aware of their expertise, power, usefulness, responsibilities and resources so as to ensure their instant contribution. Participation of women and other sections of society promote unanimity in the group functioning.

It is obvious from past experiences that in vulnerability analysis and risk assessments all the needed information is not provided by the specialists. It is essential to supplement expert analysis, usually given by the experts such as economists and scientist through a vulnerability analysis conducted at the community level involving the people. Vulnerabilities are susceptibilities to risk in the community should be analysed in relation to their capacities and resilience power. Hence assessment is a participatory process undertaken in phases involve data collection, interpretation and analysis of both scientific and empirical data.

Advantages of community based Risk Assessment

If there is a difference of opinion between the public and the planners then the planning has failed to make changes in land use that are desirable.

Studies in different parts of the world reveals that community in disaster stricken areas knew both the risks and the remedies, but the political will not apply the remedies until the disaster occur. Vietnam has developed a grass root democracy programmes which call for discussion of development policies and budgets among local people along with the government departments. The people's committees are more active in formulating development plans for land use development to alleviate poverty.

Traditionally the peoples of hazard prone region of pacific islands have been known to build their houses from local, lightweight and strong materials that could absorb torrential rains, withstand to high and gushy winds of typhoons and can resist the shaking of earthquakes. Strong social networking are a key strength, which help people in disaster situations and provide psychological and social support. In Marathwada, good quality stone masonry structures are more resilient to shocks than modern structure.

Many examples from all over the world shows that if community is involved in disaster mitigation strategies then without a planned preparedness programme community respond much better if the community develops activities and initiatives based on community discussion and action to create an understanding of the causes of disasters and associated problems. So that they can plan the most appropriate measures and put them jointly into effect.

Chapter - 5

Vulnerability

Vulnerability is measured in terms of degree of loss to a given element at risk resulting from a given hazard at a specific severity level. The measurement of loss lies on the element at risk and may be measured as a ratio of the number of killed and injured to the total population as the amount of damage is always defined on an appropriate scale.

Researches shows that during an earthquake the peoples who suffers most are single parent families, women, elderly people, children, handicaps and aged are particularly vulnerable social group. (M.Erdile 1953).

Indicators of Vulnerability

Indicators of vulnerability are poverty, marginalization, access to resources and level of development

Vulnerability can be divided into:

1. Physical vulnerability.
2. Socio-economic vulnerability.

Physical vulnerability

Means vulnerability of buildings at risk, infrastructure, physical location of people and their proximity to the hazard. Technical capacity of buildings and structures to resist them from geo-physical hazards.

Socio-economic vulnerability

Social vulnerability can be divided into two aspects, individual and collective vulnerability. Individual vulnerability is calculated by the access to resources, social status of a person within the community. The collective vulnerability of a social group is determined by economic and commercial structure of the society. Penned capacity to absorb or avoid the impact of hazards, its effects are very clearly detected. Economic vulnerability measures the risk of hazards causing losses to economy of a place. It evaluates the direct loss potential, indirect loss potential and secondary effects. Through it is possible to estimate losses and to find ways to mitigate them in relation to the estimated costs of relief and recovery. For all the phases of loss assessment, the primary source of data will be information of past disasters, so there is a need for maintaining precise records. On the basis of this data, reasonable predictions for the future losses in a similar situation can be inferred. This data is used by emergency planners and for risk reduction techniques. Information from hazard maps and vulnerability analysis can be used for the evaluation of economic vulnerability.

Thus assessment of risks and vulnerability analysis is a multidiscipline task. The product of these activities is always a comprehensive assessment of the causes and effects of hazardous events and mitigation measures.

All the societies have to live with risk of various types it is impossible to remove them instead mitigation measures have to be adopted to reduce the losses. The cost of recovery must be with the cost of preparedness and preventive measures.

Vulnerability Atlas

In the pre disaster period efforts would require identifying the areas, which are disaster prone, nature and severity of disaster also calculated which are likely to hit the areas. If correct information is collected about the above aspects then it will become an important tool to evolve strategies and initiate steps to mitigate the impact of disasters by reducing the losses.

A vulnerability atlas covers details of each hazard may occur in a region includes hazard maps for areas prone to it. In India statewide maps and risk tables for all districts in the country has been prepared by Ministry of Urban Affairs. These Atlas should be circulated all the concern departments and ministries working for disaster management. These Atlas also needs to share will the educational institutions engaged in architecture, engineering, planning and students also make aware of it.

Vulnerability maps are very useful tool to all planning departments, the hazard prone areas should be safely avoided for important uses the losses can be reduced. These areas can be used as open and green lands and recreational places.

Vulnerability Analysis

The first and foremost issue is that of vulnerability analysis should be carried out in detail. This will provide an idea about type of element, location and numbers which will help in planning of the measures to be taken in the three stages of the disaster management.

Identification of vulnerable sections of the roads, low lying areas, old houses, assets likely to get affected and so on.

The information collected helps the decision makers, planners, and officials to identify the resources, process to be adopted and help in the preparation of preparedness plan.

It can identify the short comings of actions and contingency in the preparedness plan. It also help in verifying the modification of the available data.

It can also help in creation of resources for stable and durable development by reducing the vulnerability.

Quantification of Vulnerability

The vulnerability of an element is usually expressed as percentage of loss for a given hazard severity level. Vulnerability for a range of events of different severities can be given by means of a Damage Probability Matrix. Where data is available, a continuous vulnerability function mapping values of damage to values of hazard severity can be defined graphically or numerically as an equation.

Vulnerability of structure is based on the building type, Construction materials, Foundation type and elevation Location of the place, wind load capacity etc.

Fragile ecosystems like wetlands etc. must be understood for potential hazards such as oil spills, chemical hazards etc.

Vulnerability reduction

After disaster risk reduction vulnerabilities are identified and finally actions are taken to reduce vulnerability. The most powerful defense against disaster is a society that is generally less vulnerable resistant to hazardous impacts. The reasons for vulnerability of communities are different for any type of disaster and it is not possible to tackle the root causes of vulnerability along with that the strength of that society will be improved.

The most vulnerable people of society are economically weak people. The worst locations rugged slopes prone to landslides, the low lying floodplains are the most vulnerable places to live.

Vulnerability is directly related with the levels of development in an area and its reduction needs to address within the developmental context. Protection of the resources and improvements in the economic potentials of a society can be as critical as reducing their vulnerability.

Parry et al. (1985) notes that it is necessary to include economic and social vulnerability and the sensitivity of the physical system when studying vulnerability.

The relationship between technology and vulnerability can be analyzed from biophysical vulnerability to environmental change because irrigation, improved seeds and fertilizers will mitigate rainfall deficit. Dams and other engineering efforts prevent flooding and new and old technologies provide alternative livelihoods for those living in degraded landscapes (Slater and Levin 1981).

Approaches to Vulnerability Analysis

Echno-Centric Analysis

Vulnerability is defined in terms of threat to the elements at risk such as structural vulnerability of buildings, bridges, health infrastructure etc. Faults and weaknesses in the design are analysed, Soil type in the basement, active faults, location analysis, possible alternate sites are also studied in this analysis. It involves use of mathematical models for risk assessment of physical infrastructure. Loss estimation on the basis of data collected for risk assessment.

Target Group Analysis

Vulnerability is analysed in terms of social, cultural and environmental aspects.

Planning for disaster

Disasters natural or man-made need proper planning to control disaster preparedness and prevention are very important tasks to control disasters. Human errors involved in the failure of systems which lead to accidents can be reduced through proper planning. We can minimize losses both to human life and properties but cannot stop them so our focus is on limiting the loss as much as possible, which is the basic concern of disaster planning.

An underlying problem with the emergency planning is that the fundamental nature of disasters is that they cannot be predicted with hundred per cent accuracy. So the best alternative is to prepare at all times so that whenever disasters occur we can counter them quickly.

Disaster Planning: Planning for disaster starts with prediction of risk of an event like floods, earthquakes, hurricanes, fire, spill etc. and possible effects on human life, property and the ecosystem. These factors are very important while taking into account the planning for disaster.

Requirements for the response, protection needs, rescue and evacuation, relief and rehabilitation are ascertain in pre-disaster period. The areas with highest risks are determined. The planning can be done quickly and satisfactorily with remote sensing and GIS. Disaster planning is very effective and powerful when modeling is incorporated into GIS. Modeling allows a disaster manager to view the probability of occurrence of a disaster, the point of maximum impact and damage, locations where life and property are at the highest risk and what type of response resources are required etc.

Immediately after the occurrence of a disaster, first task is locating disaster assistance centers established after a disaster has occurred, availability of shelters is mandatory on the basis of number of victims.

For effective preparedness, communication and training for disaster management a perfect plan is needed.

Disaster contingency plans

These plans are normally focused on how to address hazards particularly. Its focus is also on mitigation and recovery measures but it is not for entire disaster management cycle, such as rehabilitation and reconstruction are not included. An effective national planning will usually generate various contingency plans to meet specific disaster conditions.

Disaster preparedness

Disaster preparedness is a pre disaster phase in which we try to prepare yourself for disaster, it involves forecasting of disasters and taking precautionary measures prior to an emergency when advance warnings are there.

Preparedness includes the regular testing of warning systems and evacuation plans and other measures to be taken to reduce potential loss of life and physical damages in addition it also involves the education and training of personnel and population at risk.

Capacity development covers strengthening of institutions, mechanism and capacities of all stakeholders at all levels. All the national, regional and nodal level building institutions and resource centres at district and state level.

Education and training must be included inclusion of disaster management in the curriculum of universities and schools. Training of disaster managers, drills and other activities to make aware people about how to response disasters. Mapping of vulnerable groups.

Disaster preparedness plan

Planning with reference to disaster preparedness is holistic. It proves clarity in monitoring, control and direction which are require din disaster crisis. The plan should be action oriented involving a wide range of people and organizations. The purpose of planning is to anticipate future conditions and requirements.

A Disaster Preparedness plan: Provides data for hazard identification it gives information about the resources, demographical profile, and existing institutional set up, at all levels.

It also defines specific roles and responsibilities for various actors. It ensures networking with media agencies, NGOs, international agencies and other stakeholders.

The plan needs to be monitored from time to time and updated.

Participation of NGOs in Disaster preparedness Plan

The government alone cannot meet the needs of the people in such crisis. They should take the help of NGOs in their work.

NGOs can help in specific tasks assigned to them. NGOs with proven track record can contribute a lot as they work with dedication and without self-interest. NGOs can easily penetrate the community and create awareness among the people.

Role of Remote Sensing and GIS in disaster management

Remote sensing is a source of data collection about an object form a distance without touching it. Remote sensing of the environment is done with the help of a device known as remote sensor. These sensors are mounted on aircrafts and satellites. Remote sensing sensors use visible and infrared portion of electro-magnetic radiation (EMR) for collection of information in addition to it there are remote sensing satellites which are collecting information about the earth day and night through the use of radars and multispectral scanners.

Remote sensing yields two types of data products – black and white aerial photographs and FCC (false colour composite) imageries.

Remote sensing imageries have applications in mapping land use and land cover, agriculture, soil mapping, foresting, urban planning, disaster management, hazard mapping, hazard monitoring, etc.

Gamma rays are entirely absorbed by the earth's atmosphere and are not available for remote sensing.

X-rays are entirely absorbed by the atmosphere and are also not available for remote sensing.

Ultraviolet rays are absorbed by the ozone layer and are not available for remote sensing.

Photographic available for remote sensing the earth.

Visible available for remote sensing the earth.

Infrared available for remote sensing the earth.

Remote sensing data is widely utilized to conduct multi-hazard mapping, monitoring of floods, cyclones, fires, oil spills, earthquakes, etc. The 2004 Indian Ocean Tsunami effected areas were mapped. Pre-disaster remote sensing image database is indispensable (Jensen and Hudgson, 2006). A high resolution data of multi-time period is required for hazard monitoring.

When disaster strikes, high resolution (0.25-2m) panchromatic or near infrared data should be acquired within 12 hours to 2 days (Schnitzler and McLeod, 1999). By comparing pictures of pre-disaster with post-disaster images manual and digital change detection takes place (Jensen 2005). To assess damages and allocating scarce resources 0.25-1 m panchromatic near infrared data is required for quantitative information about damaged housing stock, spilled materials, etc.

GIS data bases finds its applications in many areas such as cartography, surveying, photogrammetry, remote sensing, landscape planning and geo-statistics. GIS is a computer based data management technique that is used to capture, store, retrieve, analyze and display geographic data. GIS data has both spatial and non-spatial attributes.

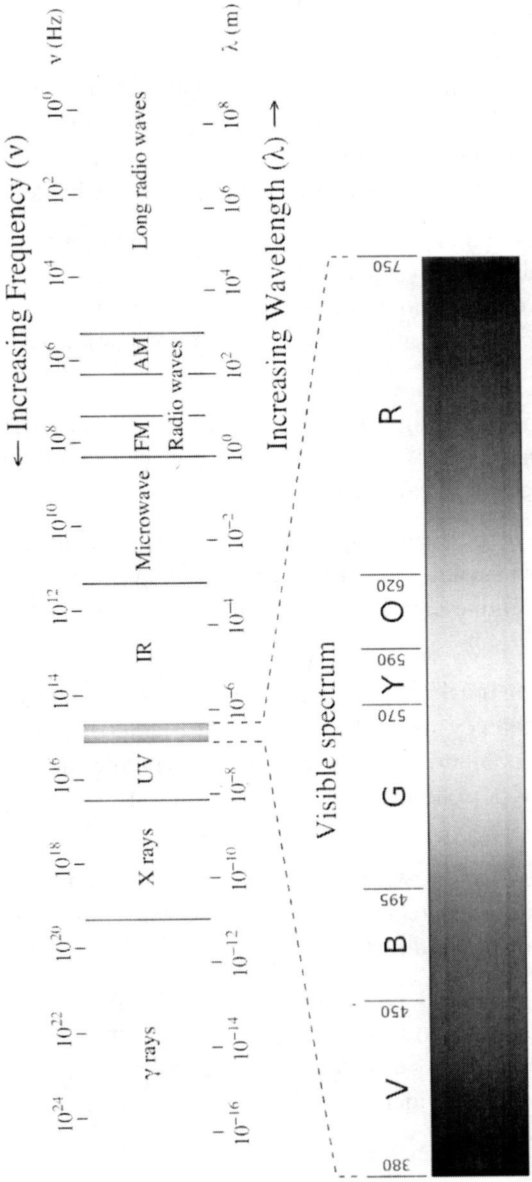

Source: Wikipedia.

Fig. 21: Electro Magnetic Spectrum

GIS is an effective tool to convey and disseminate information for mitigating disasters through monitoring relief assistance and rehabilitation.

Remote sensing can play a vital role in flood hazard monitoring and it can be used for information integration by GIS. Floods can be easily monitored, mapped and flood hazard zoning and post-disaster survey of rivers can be easily done by remote sensing technology.

Remote sensing and GIS technology can easily be used in the management of droughts by early warnings of drought. Satellite data can provide knowledge about potential groundwater sites. Remote sensing can also be used satisfactorily used for evaluation of areas subject to evaluation.

GIS and remote sensing can also be used for the monitoring of fault lines, preparing of seismic hazard maps to evaluate earthquake hazards and try to reduce earthquake risk.

Demarcation of areas with slope stability and instability to prepare landslide zones, a landslide hazard zonation map helps in the location and extent of the slope area likely to be affected and rate of mass movement in an area.

GIS can help in quick analysis of a disease spread, preparation of flow charts in emergency management of post disaster relief and health services. The vulnerability records of various human produced disasters such as leakage of poisonous gases, risks, bomb blasts, etc. can best be monitored by GIS.

GIS can be used in disaster preparedness plan GIS technology is an effective management tool for government agencies to improve their efficiency during the disaster.

Minimizing of losses, protection of people, property and environment can be achieved through the implementation of emergency plan. Routes of evacuation during earthquakes, floods, cyclones etc. can easily be identified. The GIS can play a crucial role in the mitigation of disasters.

Chapter - 6

Logistic Management

Logistic management is directly associated with material necessities planning. It includes stocks and requirements. Distribution of commodities to victims at the disaster location is very difficult. The assignment of distribution is commonly given to a distinct agencies.

Logistics Support Plan

Transport facilities are the major factor that play a crucial role in the provision of supplies to the respective locations, which includes the exchange of materials between production facilities, warehouses, distribution centers, ports and affected location. Transport structure must be resilient enough of disasters so that commodities can be provided to victims. However, time is the most important factor to provide the needs to the people. To manage disaster relief supplies it is important that we have complete knowledge of warehouse and storage facilities. The storage sites should have a planned and organized delivery system.

Principles of Logistics Management

A logistic plan should be able to support disaster operations in the post disaster period and support the efforts in the long run.

Inventories are made and goods are supplied much before the disaster occurs. Inventories are stocked at zonal earmarked warehouses and updated on regular basis.

Type of disaster must be taken into account as the needs vary according to the disaster

Locally available resources can be utilized as they are cheaper and available locally.

Logistics in the management of disaster

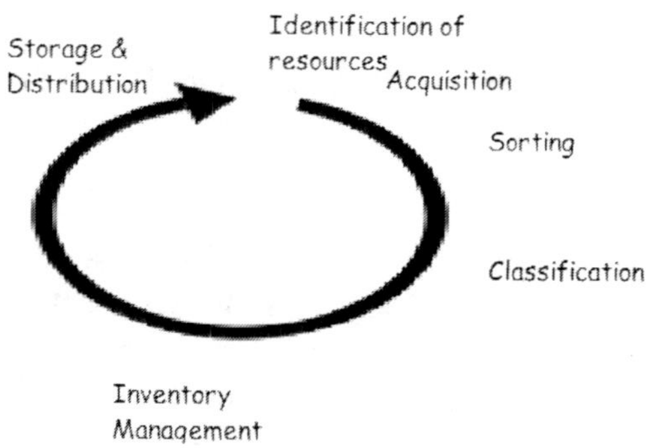

Fig. 22: Logistics Support Plan.

In the pre-disaster period goods are stored in warehouses for use during disaster it is a part of disaster preparedness, these stored items supplied to the disaster sites immediately after a disaster strike to time lag can be reduced.

Hence, in every phase of disaster management cycle. During a disaster management agencies work with government and NGOs with the emphasis on logistics and supply chain that involves the logistics personnels. All these together create humanitarian logistics stream to reduce the impact of the disaster to provide the victims with urgent needs (Thomas, 2001).

Logistics management during disaster

To manage and execute the logistic support during disaster is very tough. In most of the cases the communication system and the infrastructure are severely disrupted having an equally disruptive effect on logistics supplies.

Proper logistics planning is required for the successful management of the disasters to ensure timely delivery of the needs.

Elements of logistic management

(1) Quick response towards the disaster is the most important aspect. An efficient and suitable transport system must be available at the site to carry on the logistics operations. Following tasks must be done properly –

(i) The proper relocation of resources and management after warnings are issued.

(ii) Evacuation plans

(iii) Assessment

(iv) Road clearance for the movement of the facilities

(v) Moving rescue and repair teams (UNDP, 2006: FLIS 2010)

(vi) Transport: Mostly the relief operations use roads and air transport. During the relief operation the co-operative freight systems for the transfer of resources must be done. It enhances the quality of the operation.(Aitkens et.al, 2002, UNDP, 2006, Iqbal et.al, 2007).

(vii) Co-ordination and collaboration: A good co-ordination among the agencies involved in the process is required for successful accomplishment of the task, governments should play authoritative role in determining the critical resources victims need a good co-ordination among the NGOs and other organizations is also required for the logistic support.

Protecting backup Stocks

Once a disaster hits an area such as earthquake the aftershocks occur frequently, therefore, proper management to store the supplies in the disaster area must be done to achieve the goal of providing aid to the victims (Ozdamar et al, 2004: NDMA 2010).

Monitoring and Evaluation of Logistic Support

The logistics support plan needs monitoring and evaluation for its proper implementation, at the local level. The monitoring involves the date and time of consignment shipment and delivery must be informed to the staff, the means of communication, radios, transistors etc. Must be properly structured. Maintenance of record of dispatch of goods from each warehouse, most important is the supervision of the delivery to final beneficiaries.(UNDP, 2006.FEMA, 2013). For the evaluation of logistics, assessment team must work in co-ordination to support the administration. The information required for the assessment of the system are destruction of roads and bridges cause hindrance. Forecast of weather conditions etc. for monitoring and assessment needs trained management staff to achieve the goal.(UNDP, 2006).

Geo spatial technology is widely used nowadays for efficient decision making. This technology includes remote sensing, Geographical information system, and geographical navigation satellite system (Karen et al, 2009).

Stress Management

A disaster is a severe disruption, ecological and psycho-social, which greatly exceeds the coping capacity of the impacted, (WHO).

After a disaster there are many who left in grief and sorrow because they lost their dear ones or fell in serious injury which effect on their lives so badly, individuals level up preparedness is therefore most importance. Thus training and education are crucial determinants of a positive outcome commonly after disaster the victims show numbness, traversing elation, anxiety, flashbacks, and night mares.

Psychological morbidity tends to effect 30-40 per cent of the population in first year. After two years, percentage are lower, but in some individuals it persist and become chronic. Human induced disasters with high shock and destruction show severe impairments physical disorders and mortality rates are higher for survivors of disaster, who have experienced severe physical and emotional trauma, material help alone is not sufficient. Alleviating the trauma and helping people to reclaim their lives is essentials.

Psycho- Social Trauma

The prevention of trauma is very difficult. However occurrence of particular responses such as post disaster stress can be diminished, if not avoided.

Programs organized for Trauma relief they teach participants how to control their emotions and stress through breath, also efforts are made to shift their attention away by ignoring the past through future possibilities. Participants are forced to attend the workshop on stress and trauma management techniques, they learn through these programmes coping strategies, a stronger community bonds are build and a sense of personal power is created.

True relief is served when disaster survivors are fully rehabilitated physically and emotionally. Natural disaster can cause negative effects on health and emotional wellbeing on those who lost their belongings and family members in the disaster reflecting back on it. Survivors of Natural disaster, such as the 2004 tsunami, suffered psychological stress that will harm their mental health and there are many type of risks, symptoms, emotional responses that are associated with this distress. Some survivors experience long term psychological stress impaired mental health is also due to the impact of the disaster such as physical injury, the loss of loved ones and the perception of life threat.

Post-traumatic stress intense or unpredictable feelings- anxious, nervous, grief-stricken. Changes to thoughts and behaviors sensitivity to environment can be created Patient's education, social support and anxiety management through physiotherapy, education and social support are from family and friends emerges understanding and acceptance.

Minimize media exposure

During a traumatic event or disaster it's important to protect the victims from unnecessary exposure to additional trauma and continuous reminders of traumatic events.

Distress and trauma are normal behavior in post disaster efforts should be made to minimize stress that may set in after the event. Shock and denial are common responses to disasters, shortly after the event time should be given to victims for healing. Ask for support to community and relatives who are near the victims and who will give attention to victims and communicate with them.

Disaster Training

To handle the disaster related physiological problems health workers must be trained before working with disaster victims. Training programs should include first aid, medics, nurses, teachers, social workers, etc.

Administrators can be trained to identify vulnerable group.

Disaster training varies from one country to another depending on the requirements of each country such requirements may be the size and geographical location of a country, a disaster prevalent in a country, the availability of disaster related resources degree of dependence on international assistance, and training capability of the personnel

Thus it is important that countries should carefully examine, their individual circumstances and then establish their training requirements, and they must know how these requirements will be met.

Why Training

Disaster management includes various activities therefore training is required for planning, organization, management, and counter disaster operations. Crisis management, logistic management, recovery and involvement in national development programs, preparedness and prevention.

Training is required for members of government agencies, non-government organization and community volunteers.

The main objective of training is to build the capacity of civil society, disaster managers and volunteers, and foreign delegates, for better preparedness and response at all levels and all phases of disaster management to work together in a coordinated manner. The main objective of disaster training is generally to improve the technical skills of the participants and it will also create personnel and team management, it helps in the exchange of experience and knowledge and networking is created amongst the disaster managers. Through it government can improve coordination of disaster response and the quality of disaster management tools and techniques.

Training is supported by government of the countries as per the requirements. This training will reflect the involvement of the national society in disaster management in its country and vary greatly.

Disaster Training includes the following:

- Induction of Volunteer
- Volunteer management techniques.
- Community based first aid and trauma management
- Community based disaster management (CBDM).
- Hygiene and sanitation to improve health conditions.
- Water and sanitation during and post disaster period.
- Vulnerability and capacity assessment needs assessment training.
- Relief distribution and Logistic management.
- National disaster response team.
- Public health emergencies Food security and nutrition.
- Telecommunications.

Disaster management training at regional level includes

Regional disaster response team (RDRT) or Regional intervention team (RIT)

Regional disaster response team leader training Food security and nutrition, relief, vulnerability and capacity assessment, emergency needs assessment.

Disaster management training at international level includes:

(1) Field assessment

(2) Emergency response team

(3) Team leader

(4) Relief

(5) Emergency needs assessment.

Disaster warnings and forecasting

Natural disasters can be predicted up to considerable extent, predictability helps to understand the nature of a disaster, assess the chances of occurrence and possible destruction by the disaster. Forecasting and warnings are a crucial part of a disaster must be based on sound scientific principles and techniques, therefore forecasting is generally done by authorized agency or individuals who are comfort, qualified, responsible persons. A successful disaster management depends on a successful prediction and forecasting. However, quick transmission of correct information is. Catastrophes are growing by the decade, global annual disaster costs of fifty billions US dollars approx. Between 1960 and 1990 the economically losses by disasters increased five times due to rising vulnerability (Munich Re, Nat Cat Services. Geo Risks Research, 2005), mandatory for an effective disaster management.

Floods and droughts, cyclones and cloud burst can be predicted monitored to a great extent thus they are predicted correctly.

Floods forecasting is easy all over the world as we can easily monitor the rivers. However flash floods, water logging due to inadequate drainage forecasting can be done locally.

If peoples are involve in the dissemination of early warnings then it will empower communities and prepare people to face the natural calamities. The efficiency of a warning system is calculated in terms of lives saved and losses are reduced which is directly related with the response of the people and institutions after a warning is issued.

The phases of Early Warnings

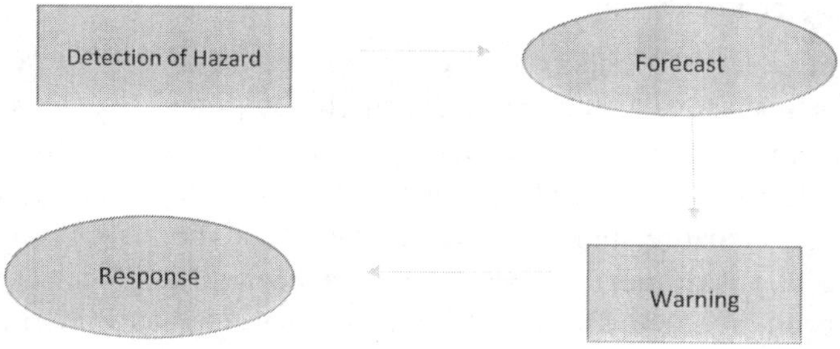

Source: Created by the author.

Fig. 23: People Killed by lightening .

The figure shows different phases of early warnings.

Different hazards required different type of early warning systems.

In 2004 more than 1 lakh people due to lack of a tsunami warning system in the Indian Ocean. Countries of India, Sri Lanka, Indonesia etc. The lack of awareness, lack of training to understand a warning. The early warning is composed of phases:

1. Identification of hazard.
2. Forecast of an even.t
3. Release of an alert or warning
4. Response to a warning

Challenges

1. Different hazards require different early warning systems.
2. Early warning systems are placed at risk points, but problem is that there is a weak coordination between the warnings issued and people's behavior to respond effectively to the warnings.
3. Many countries have limited facilities for effective early warnings system. Hazards and vulnerabilities together can be taken to reduce risk, it is important to increase the effectiveness within institutions and peoples preparedness for early warning systems to make them more effective.

Towards Effective Disaster Preparedness

1. Disaster Leadership

The concept of leadership in disaster response is very crucial as disaster management requires planned strategies and people with leadership qualities to accomplish the tasks. According to Foster et al, (2012, 129 -133), Leadership in disaster and emergency response is that of mutual relationship between the leader and his followers. The relationship here, is considered to be interdependent and reciprocal. In the situation of catastrophic events disaster leaders are a powerful source to handle various tasks such as distribution of disaster relief and rehabilitation, at various levels, disaster leadership can be categories at following levels:

(1) Local Leadership

(2) State Leadership

(3) National Leadership

(4) Regional Leadership

(5) International leadership

Leadership may be defined as quality to influence or motivate people of a community for achievement of certain goals. In general leadership is entirely different as decisions are taken, after enough considerations as well as under the thoughtful guidance of experts. In the emergency situations or during disasters, tasks of leaders usually becomes difficult.

There are various factors which affect leadership from local level to national level they are:

1. The local leaders living in the affected areas can also themselves caught by the disaster. They could get isolated due to sudden breakdown of communications or become ineffective under traumatic condition due to the suddenness and severity of the situation during disaster, many of the relatives and friends of the leaders could also be affected. Their attention could get diverted to their problems, instead of taking decision or action for relief and recovery of the other members of the community.

2. Lack of information and disconnection of communication become serious factors for decision making.
3. Loss or delayed supply of resources, equipment, transport and other relief material delays actions and create a sense of helplessness. Resulting in the confusion, the community feels insecure and could lose confidence in the leadership. There may be some other factors also depending on the type of disaster and the affected people.
4. In a crisis situation requiring relief and rehabilitation of disaster affected people, administration and political leadership could have different goals.

(A) District administration

At the district level the district magistrate or collector is chairman of the disaster relief committee. The main objective is to organize effective rescue operations for the people caught in the disaster situation.

Another objective of district administration is to provide immediate relief like food, medicine, drinking water, clothes and temporary shelter. It depends on the leadership qualities of an administrator, how they finish the work.

(B) Local leadership

Local leaders must be available at the right time during emergency either they are experienced or not for working disaster management but in some cases, their intervention may hinder the process of relief but may give political advantage to the local leaders.

(C) National and State level leadership

Leadership at the State and national level can provide funds and resources and mobilize effective relief within the available resources but this leadership initiative from top will be based on the political will and the disaster situation. To make the leadership effective, it is essential that they should have accurate and up to date information from the disaster area through the district leaders and leave the task to the local administrative system for implementation.

LEADERSHIP STYLES

Before dealing with the leadership styles, it is important to know the attributes and desirable qualities of a leader. Leadership qualities cannot be learnt from a book; for example, 'courage' cannot be learnt from anywhere except perhaps from worthy role models. Secondly, no leader can be an all-rounder or ideal in all aspects. There are some good qualities that every person has in some measure, such as sense of humour, endurance, cheerfulness, dedication, enthusiasm, courage, quick decision, identification of problem, etc. But combinations of a large number of these desirable qualities in a person can make him or her a better leader. Some of the more desirable qualities of leadership in disaster management are briefly discussed below:

(1) Personal qualities and self-confidence

As mentioned above, every person has some leadership qualities. One should identify them and try to develop and upgrade them to the best capability. Self-confidence is an important characteristic of a leader which can be developed by increasing his own professional competence and inter-personal abilities.

(2) Professional competence

The professional competence means one should be sure shot about the style of working that how a work can be done what type of knowledge, skill and ability is required.

(3) Sound judgment and appropriate decision making

A leader should have skills of taking right decisions at appropriate time, a leader with these qualities will emerge successful with his team of in disaster situations.

(4) Ability to communicate

The communication skill of a leader is key to success for proper functioning as clear and concise communication with people working with the leader is essential, because it can help in the development of good interpersonal relations that generate goodwill and loyalty to the leader leading to a high level of discipline in the team.

(5) Appropriate style of leadership

Always different tasks require different styles of leadership. It is important for a successful leader to understand the dimensions and requirements of the given task and adjust the leadership style to achieve the desired results.

1. Orders: The leader orders the team and waits for results and action while keeping an eye on the progress. He does not expect his subordinates to ask questions or give suggestions. This style is adopted when the matter is urgent and there is no time to lose.

2. Consult: Convinces the team about the decision taken by analyzing the positive and negative points. It is like selling ideas to the team members. Obviously, this exercise takes some time.

3. Consults : Leader consults the team and allows them to participate in the decision making with the view to making small adjustments but the main thrust of the decisions of the leader are more or less final. This style is somewhat less time consuming.

4. Joins: The leader discusses thoroughly the problem and the likely course of action. Every member of the team is involved in the discussion and finally the consensus decision prevails. This style consumes most time but later work is smooth.

Sometimes a multi-style or mixed-style approach is more useful in disaster situation. There are various different tasks such as distribution of food, medicines, temporary shelter, and rescue work. If each of such tasks is assigned to a different person instead of all tasks to one, there can be better results. In other words, delegation of authority and work is also an important aspect of leadership style. Of course, the leader has to continue to monitor and coordinate with his colleagues.

Another useful factor in the development of disaster management is the strong and positive link between leadership and training. A high standard of training can upgrade the professionalism in the leadership. This is the reason that the Government of India (through its nodal ministries and departments) and the various institutes of public administration (through their faculties of disaster management are giving more emphasis on 'Human Resource Development' in the area of disaster management. They want to provide training to various levels of government officers, NGOs and to the community leaders so that they have more knowledge, skill and confidence to tackle the likely disaster situations.

Different type and scales of disaster raise varying aspects and level of disaster leadership and their responsibilities

Disaster leaders and managers should be aware of these and get ready to handle the resultant problems.

A disaster leader must have these qualities like personal traits and self-confidence, professional competence accurate decision making and sound judgment, good communication skill, ordering, selling consulting, participating joining and team making.

In a disaster situation always a multidisciplinary disaster leadership is more useful.

A high standard of training can upgrade the leadership.

ROLES OF LEADER AND COORDINATOR:

A leader's task is to put efforts to make the work simple and more effective. He serve the affected people within the available resources and be cost-effective. Role of a leader starts from pre-disaster phase. He has a crucial role in during and after the disaster phases.

Role of leader and coordinator are given below:

1. Foremost task of a leader is the Identification of safe places and protected areas, in a disaster situation. He can inform and make the community ready to shift to these safer places after the warnings are issued normally, people do not want to leave their houses and valuables even after warnings and even police intervention but a good leader can convince them to move to safe places. In the disasters like cyclones and floods, such timely evacuation of people before the disaster strike can save many human lives and cattles.
2. Leaders decide regarding the rescue operations in post disaster period, temporary evacuation, shelters are the

immediately required after a disaster occur and are crucial to the livelihood of the affected people. Leaders can convince people for self-help strategies and enhance spirit of cooperation. They take decisions to organize external assistance for victims who have lost their home and means of livelihood by involving people and community in the decision making, implementation of plans and their participation at every step of relief or rehabilitation process in post disaster period by keeping complete transparency.

A disaster leader can take decisions about the priorities in terms of rescue, evacuation, temporary shelters, and relief. Immediate need in context of saving life and protecting livelihoods, they can also implement self-help measurers and create spirit of co-operation, can take decisions to obtain external assistance, they can also take part in country's decision making and implementation of plans for relief and rehabilitation.

2. Co-Ordination

Disaster co-ordination is an important aspect of disaster management. Disaster management involves a very efficient co-ordination among the government, non-government agencies and community. A proper co-ordination of disaster response operations among various bodies included in it, is a key to successful disaster mitigation, relief and rehabilitation. Such a shared initiative of various agencies and departments is quite helpful in achieving the goals of the social task of disaster management, previous experience of disaster preparedness and management suggest that there is wide scope for improvements in co- ordinated activities, particularly during rescue, relief and rehabilitation. However, the main principle of disaster co-ordination which need to followed during disaster management process, which includes networking, role allocation, co-ordination during exercise assessment of available resources, finance.

A good co-ordination is key to success which make the task of disaster management a systematic and more effective, even when the resources are scarce. A good disaster co-ordination can help the safe place for shelter during disaster. The community can be successfully respond to different warnings and evacuation process should be safe and smooth.

Disaster coordination is the key to an effective response

We need a review many of the lessons we learned from the past disasters particularly those we learned in the field of humanitarian assistance, specifically in the health related issues. New agencies are emerged in the emergency situations that focus on providing immediate health services, often for short periods. Now there is a need to evolve a platform where the stakeholders interact and collaborate for major tasks study various scenarios to develop models of to apply their efforts and resources more effectively, to solve potential problems and avoid repeating the same mistakes again.

When a disaster took place and causes great devastation people suffer severely, sympathy is created with the victims, the entire world turns its attention to the disaster victims and try to serve them in its maximum power to mitigate the impact through an instant response. Humanitarian teams are mobilized and donation campaigns begin, systematically. As a result occasionally a chaos is created many type of medicines and food items are collected but due to improper arrangements go stray give rise to headache for those responsible for receiving and disposing of them. In this situation an intelligent manager can play a vital role in the organization and distribution of relief to the victims.

Efforts are made by the countries in building up their emergency response, but the miss management that often occur in the wake of disasters cannot be controlled some time. The coordinating agencies have responsibilities to manage efforts properly and respond to needs in order to ensure ongoing attention and the recovery of essential services.

The governments of the affected countries, both at national and local level, are sometimes excluded from the early phases of the emergency response, but when assistance tails off (usually after media attention has died down), they have to resume control of the situation under very difficult circumstances.

This lack of coordination in the management of humanitarian assistance can be avoided through proper response planning. If each one acts according to its own protocols, without coordination contradictions will result. This lack of coordination will hinder a proper response to the needs of the affected population.

An enormous hurdle. An initial large-scale mobilization creates expectations in the affected communities, but then the area gets less and less attention until, finally, it is forgotten.

Guidelines to ensure a coordinated response

The government of affected country by an emergency is responsible for organizing, coordinating, and providing humanitarian assistance in its territory. This coordination should be spearheaded by the national authorities, which should provide the means and the tools to properly plan and prepare a response based health action plan for the victims. Governments are responsible for maintaining response teams, with organized procedures for receiving and distributing humanitarian assistance, so that they will be able to respond appropriately and effectively in their role as coordinators.

The institutional response to emergencies and policies for disasters are prepared on the basis of a number of existing disaster-management models. Internal and external evaluations of the experiences of different organizations like WHO in the emergency response are mandatory.

To ensure an effective response, it is important to emphasize the leadership role played by individual countries through their respective ministries of health, as disaster response coordinators. On the other hand health ministries must be strengthened and seek strategic partners who will unite their efforts around the response plan proposed by the authorities, thereby achieving better health coverage, better use of resources, sustained action over time, and greater cost-effectiveness at the time of response.

Disaster communication

Disaster communication includes a wide range of actions, techniques, ways and models of communication, to impart knowledge and information, warnings and forecasting of disasters threats. Significance of communication is a key to every phase of disaster management, particularly in disaster mitigation.

Techniques of communication:

Major areas where communication is required includes:

1 Education and literacy
2. Public awareness generation programs
3. Training and industrial manpower
4. Media information
5. Transmission through wireless sets and radios
6. Telephones and mobiles
7. Online data sharing through net

The above communication techniques are equipped with new technology and available everywhere so that they are successfully used, all these ways are very helpful in communication.

The principle ways which ensure effective communication include a proper planning of requirements, a directory of disaster management, skilled manpower and further capacity building vigorous and continuous updating and training, well executed satellite control stations and expenditure sharing.

It is a worldwide fact that in a disaster situation, every second is very important. Any delay in communication and disaster response could become cause of serious aftermaths.

Disaster and communication

Natural disasters are integral part of environment, responsible for loss of life and property.

Communication is an important component of response and recovery especially during and immediately after a disaster situation as it connects affected people, and communities with first responders and support systems. A reliable and accessible communication system have key role to a community's resilience. The communication technology has been considered as integral part of disaster management. Although communication technology has a crucial role in all the phases of disaster management, but most important application has traditionally been in response and recovery. The new communication and information technologies emerged over the last two decades had improved the communication system greatly. Various communication systems including internet, mobile phones, fax, e-mail, radio and television are inter related and this characteristic make them more relevant for disaster communication. As a result, the possibilities for application of communication technologies in mitigation and prevention of disasters are also increasing, all disaster emergencies and crisis events are chaotic and highly dynamic.

In such crisis events and emergencies, communication is required at all phases of disaster management. Means of communication during emergencies incorporates a wide range of measures to manage risks to save communities and the environment, before disasters strike, after the release of early warnings the telecommunications can be used to disseminate information about the impending danger thus, people at risk can be made aware and necessary steps can be taken, evacuation can be done for vulnerable areas making it possible for people to take the necessary precautions to mitigate the impact of these hazards. Other telecommunication systems, including remote sensing and global positioning system (GPS), have critical roles in tracking hazards, issuing warnings to the vulnerable population, coordination for relief operations, assessment of damages and provide support for rehabilitation and reconstruction. Now it is clear that role of communication can play a pivotal role in disaster prevention, mitigation, and response. At the time of natural calamities usually cell phones, and landlines etc might not work or might be lost in the calamity. The society at large must be well protected, this protection must be given by the police or disaster management force immediately after the strike. The disaster managers must be well trained and equipped with the best appliances which might help them to communicate the condition they are in and the status of the calamity in the region they work for.

Communication and Technologies

The disaster management communication systems must focus on their use in remote parts and even inaccessible areas like deep waters and mountain tops. So many people getting lost while crossing the forests or mountains during disasters. Many accidents occur during avalanche and landslides, so a communication system must be reliable and must be in working conditions at all times. A wide area network with satellite signals is usually helpful. New discoveries and easy to use communication gadgets are helpful in efficient communication, previously communication with remote and inaccessible areas was a great challenge but with the introduction of satellite communication, the problem of connectivity to remote and inaccessible areas has been resolved now. Satellite communication provides the facility of coverage of both national and international areas. Low Earth Orbit (LEO) satellites are orbit around earth at a distance of few hundred kilometers providing the facilities of remote sensing. Satellite radios can also plays an important role in disaster management. Arial photos and imageries from the satellites can be used to show the severity of the disaster and they can almost penetrate closer to any region to get a good view. Communication has made this world very small and transparent wherein we can see the storm ravaging in any part of the world live on our television sets. Wireless communication devices are popular these days as they use satellite signals for communication. Policy and Institutional Framework for Emergency Communications Policy and regulations are important elements of emergency communications planning and management. One of the main hindrances to effective establishment of communication system for disaster mitigation is the lack of a legal and regulatory policies. Cooperation among specialized services at each level is as essential as the vertical lines of communication for emergency communications. This requires established links between disaster coordinators, telecommunications authorities and service providers at each level. Governments must take important steps to eliminate the monopoly of public telecommunications enterprise and to undertake a review of current telecommunication legislation, including regulations incorporating emergency telecommunication systems and protocols to support emergency services.

Challenges

An understanding of basic principles of effective early warning systems is required for the effective application of emerging communication technologies, they can be satisfactorily used for a sound early warning system.

There are disparities in communication infrastructure among the Asian countries as well as across the countries of the world and among different cultural groups. The scientific and technical community is very enthusiastically making use of emerging communication methods to share real-time information as well as local knowledge and experiences, the decision makers who have key role in managing disasters have to rely on rather conventional means of communication. Communication network for disaster management is inspired by military operations, which follow a well-defined command-and control structure. However, growing emphasis on disaster management at community level and greater emphasis on the effectiveness of community-based disaster management would require that the community looks into new approaches for the application of emerging communication technologies in disaster management.

The effective application of these technologies is not possible as language is a major barrier in the world.

Survival strategies during a Disaster

Natural disaster can strike anytime and often with little warnings. Immediately after a disaster essential services are cutoff and local disaster relief and government help may not be able to reach the victims.

Create a disaster plan

Learn about the disasters prevalent in your area. Learn whether hazardous material is produced or stored near your area, learn how to response and how to save yourself. Talk about what you need in an emergency evacuation.

Plan how your family members stay in contact if separated during emergency, identify two meeting places, the first should be away from your neighborhood.

Store sufficient water, food, medicines and other essentials if warned earlier for a disaster going to occur.

Preventive strategies are attempts to avoid the disaster altogether or reduce its scale considerably to the extent possible. This technique of survival strategy requires successful political mobilization at all the levels.

Make a food kit for basic survival. Having an advanced knowledge to sustain life through whatever is available in your surroundings is critical, not just for you but for your family and the people around you. Here are a few disaster survival skills you might find useful in the event of a man-made or natural disasters.

1. Residing at the Risk

The net provides unlimited information especially on locations where the risks are high. Choose a place with the lowest possible risk.

2. Avoid areas with high Disaster Risk

Always avoid vulnerable areas because in these areas during an emergency may be a disaster waiting to happen. Learn to identify these life threatening hot spot and steer clear from them as much as possible this is strongly recommended.

3. Keep in mind three rules

Three hours without shelter, three days without water and three weeks without food, ready to learn how to survive a disasters by facing the elements head on, keep yourself away from falling into starvation and dehydration.

4. Overcome the Drought

During a dry spell, there is scarcity of water there are other dangers that come along with it. That is why planning ahead and preparing for the lack of rain and water supply will help you preserve the life.

5. Advance planning

If one live near woodlands or forests, recognizing the dangers of wild land fire could bring to one's life, livestock and property is critical. Extra vigilance is required, especially if that area experience drought.

10. In case of fire

Always be careful while planning public buildings and private housing that in case of fire the safety rules should followed, what safety measures can save yourself and your families and other people, fire extinguishers should be there.

11. Flooding and storm surges accompanied with turbulent winds are the signature of a cyclone or hurricane. Knowing what potential dangers it could bring is a must. You'll never know when you'll get blown away.

12. Stop bleeding of an injured person first to prevent his life Learn also to improvise without the person losing a limb in the process.

13. During a flood

Overwhelming waves of water can be due to natural or man-made causes the destruction it brings. Here are a few flood survival tips in order to survive a natural disaster.

14. How to Prepare for a Tsunami

Peoples living along the coastlines should always be ready for a tsunami at all times. A thorough knowledge of causes and risks will be enable you to put every tsunami safety tips.

16. Starts from the Basics

First try to build a first aid kit. The important thing is you're taking the first step to being prepared for the worst. Having a first aid kit to start with is good.

17. Reinforce safe points in your home

By consulting with a reliable architect, we can identify certain safe points and strengthen the weak areas, to make it safer in the event of a disaster. You'll never know which one will come first so it's best to prepare for all of them.

18. Prepare for landslide hazard: During the disaster avoid the areas where land can slide confined to the house if it is in the safe place or move to the safe places.

Being prepared for whatever Mother Nature has to throw at you is essential for your survival wherever and whenever she brings it on.

Chapter - 7

Role of Media in Disaster Management

News media is an institution which provides a framework of information which originates in many forms and is subject to many changes under influences for political, professional and commercial reasons. The media makes a direct link between the people and disaster managers and plays a vital role in disseminating disaster related information to the public in every phase of disaster management. Media can educate the people about the disasters, it can easily disseminate warnings of disasters, it can gather and transmit information about vulnerable areas, it can alert government, relief organizations and the public to the specific needs.

Role of media both print and electronic becomes most important and crucial.

Media and disaster response

In disaster management correct information and right time is of vital importance. People needs warnings before disaster strike and data on causalities damage and supply skills which are needed.

As public education and early warnings can save lives. For example in 2014 cyclone phylene was going to struck on Andhra coast early warnings were released much earlier and around 9 lakh people were evacuated in three days from the danger zone and life loss was reduced to only 3-5 deaths, likewise in 1985 a cyclone hit Bangladesh early warning was released so that only 10, 000 people were died still high but very low number as compare to 1970 life loss was 300, 000 in Bangladesh.

A cyclone stuck India On Andhra coast in 1971 same story was repeated 10, 000 people died, while in the storm in the same area 13 years later 910 people died, this difference was due to early warnings connected with radios to alert people on the other hand some great disasters occur and cause huge losses like Bhopal tragedy, Tsunami 2004 as there was absence of information dissemination systems. The Tsunami stuck at Indian coast well after 4 and a half hours after causing destruction in Phuket however this information was not shared with the peoples as a result great losses were occurred but this information was not shared with the people as a result great losses of life and property occurred.

The above examples show that media and its instantaneous outreach can play key role in disaster response. Media can exert pressure on the authorities for effective response measures.

If media is not careful then they can bias the nature of the response. They can create myths about disasters.

At the time of disaster media is actually helpful. Media can mobilize relief distribution in emergency response.

Types of Media

Two major type of media that exist are:

1. Electronics Media
2. Print Media

1. **Electronic Media-** is very powerful mean of information dissemination, generally radio, Tele vision are used for this purpose.
2. **Print Media includes**- newspapers, Magazines, journals.

   Technology is developing day by day new ways are developing for acquisition of data and analysis forecasting and dissemination. New advancements in communications offer the new ways for improvement information is quickly collected both during the disaster and in the post disaster period.

The development of weather satellites which are geo – synchronous, for earth observation, the use of super computers along with it to analyses data captured has led to highly sophisticated models for tropical storm formation, providing earlier and far more reliable information, with which we can plan evacuation and other disaster mitigation tasks.

Reliable information provided by the media at right time can help people overcome any fear and fatality in disaster emergencies. If reliable information is available timely it helps in mitigation and recovery, improve solidarity and creates an atmosphere for collective efforts for sharing the humanitarian challenges posed by disasters.

Media today penetrates in every place to find out the problems all over the world. Journalists wants to report objectively a global event. Media is considered as one of the pillars of democracy.

Objectives

The broad objectives of the role of media in disaster management are:

(a) Media works for public safety in emergency.

(b) Media generates public awareness about disaster risk reduction.

(c) To recommend measures that can sustain media intervals before and after disasters.

Media can play a major role during disaster they can start help line, emergency phone numbers. They can also help in quashing rumors and prevent panic situation.

Social media has emerged as a new way in the wake of happenings, with its impressive usage there is huge potential to create awareness in the society.

Media at criticism

Media is criticized on the ground that in the desire for TRP and sensationalism, the media has deviated from ethics of journalism. After the Nepal earthquake, media was heavily criticized, during Uttrakhand floods, politicians more focused more rather than disaster response and recovery.

It is essential that ethics in journalism related reporting in disaster scenarios should be strictly followed. The survivors and affected victims deserve complete privacy, their consent must be sought and too much questions must not be asked. Thus media should always involve a humanistic angle to journalism, where journalism must give preference to humanity. Over professionalism.

Understanding the vital importance of media communication departments for internal and external co-ordination, now it is believed that the success of humanitarian efforts depends largely on relations of organization with media.

The media can play a pivotal role in changing the mindset of society for making it more productive and constructive. The electronic media has an outreach to the masses the role of media is crucial in the promotion of a better disaster management process within a region. Print media particularly newspaper has great influence on the people.

Application of Remote Sensing and GIS techniques for the management of Disasters:

Management of Coastal Disasters require an understanding of coastal processes, vulnerability of hazards, and identification of risky areas and development of mitigation techniques. Remote Sensing techniques can be effectively utilize to understand and mitigate coastal zone hazards.

(Rajawat and Nayak, 2000).

Table 12: Application of space technology in the management of Disasters

Disaster	*Data used for*	*Mitigation*
Tropical cyclone	Detection, Monitoring and Evaluation, mapping of damages.	Cyclone Warning
Storm Surge and floods	Storm surge/ flood forecasting, mapping of inundated areas and selection of relief distribution sites	Mapping of the hazard prone areas, vulnerable to storm surge and floods.
Coastal erosion	Monitoring of shoreline changes and coastal erosion.	Mapping of risk zones and prediction of shoreline changes.
Earthquakes	Detection of surface deformations, monitoring of fault lines, database preparation.	Identification of risk zones, mapping of active faults measurements of fault displacement.

Space technology plays a crucial role in the coastal disaster management in India. The Indian satellites, the INSAT series, the INSAT-3A and 3D. The satellite data is extremely helpful for disaster mitigation, prevention, monitoring, damage assessment, relief and rehabilitation of disturbed population. In addition to this ISRO has planned improved sensors in its Megha Trophiques Mission, for providing information related to weather elements like water vapour, cloud, precipitation, wind speed and radiation budget in the equatorial orbit. Storms, Cyclones and other wind related hazards can be detected by Storm warning Radars. Radar units transmit a beam towards a target and then record the reflected scattered beam of radiation. The centre of the cyclone generally appears as a cloud free circular zone, and the eye is usually seen as a small circle at the centre of the cyclone. The Doppler effect or change in frequency enables us to determine whether the target is in motion and if so, what is the direction of the motion. Tropical cyclone forecasting involves, locating the position of the cyclone, assessment of intensity of the cyclone and prediction of future movement of cyclone. For damage assessments in the post cyclone period high resolution data of IRS-1C/1D LISS-III and PAN can be used. The Satellite data is also useful for disaster prevention along the coastal areas.

Storm Surge and Tidal floods are associated with the cyclones cause heavy destruction in the coastal areas. IRS-P4 OCM data is useful to understand the impact of flooding due to super cyclone hitting the Orissa coast in October, 1999 (Nayak et al., 2001).

Earthquake prone areas and damages caused by earthquakes can be easily assessed by Remote Sensing data. The Synthetic Aperture Radar (SAR) interferometry techniques are extremely useful in detecting early deformation of the crust due to an earthquake.

Through the use of Remote Sensing and GIS technology we can develop tourist destinations very successfully particularly in the remote areas where impact of disasters minimizes the prospects of tourism, in all those areas which are prone to hazards and disasters the Remote-Sensing and GIS data bases can provide valuable information about the safer locations. The periodic applications can track and monitor the cumulative impact of Land use change, within short period of time with little manpower a large amount of data can be obtained and analyzed. The Imageries and aerial photographs permit the interpreter to intensively study an area and infer information.

In the field of recreation studies, a lot is still to be done with aerial photographs. Aerial photographs have better spatial resolutions, they can be used to make inventories of tourism places, resources and facilities. Aerial photographs can also be used to interpret tourism activities, by comparing of photographs of different time period, changes in recreation and tourism can also be traced, the tourism potential maps can be generated.

The vulnerability of tourism to risk, crisis and disaster has long been evident (Sharply, 2005).At the tourist places for the safety of tourist it is essential that tourists must know the geography of their surroundings so that at the time of onset of a disaster they must know the location of critical support services such as shelters, evacuation routes, hospitals, fire stations, police stations and airports. The disaster risk management for sustainable tourism in India and the world is a necesity that can be undertaken to reduce the vulnerability of the tourism sector to natural hazards.

Tourists at vulnerable destinations are more exposed to disasters than the local people because they are less aware of local hazards and they are less independent also unfamiliar surroundings and high priority must be placed to their safety. The exposure of tourism to natural disaster is linked with the attractiveness of many high risk exotic locations. (Murphy and Baylay) (1989).The tourists also have high risk of hijacking and terrorism(Lehman 1986).Mostly the tourist destinations have risk of natural disasters like hurricanes, cyclones, floods landslides, which can easily be monitor and predicted with remote sensing and GIS satellites.

Tourism planners need to be aware of various sources of information such as aerial photographs and satellite images are very rarely used by tourism planners. They are however, a useful source of information, reducing the time and cost of scientific assessment. So, there is a need to bringing the geographers, environmentalist and tourism planners more closely together for resort planning and disaster mitigation. Due to availability of high spatial resolution aerial photographs are highly recommended for the study of tourism resorts.

Scanned photos after importing into GIS software and then be used for the mapping of tourism resorts and facilities.

Thus disasters are very common in various areas of the world and need a mitigation strategy which can successfully reduce the impact of disasters. Remote sensing and GIS techniques have great potential for reducing the tourism disasters and development of new destinations in the disaster prone remote areas of India. Tourism disasters can be avoided with the help of disaster management techniques. Disaster planning, risk assessment, prepared.

SECTION - 2
Disasters in India

Disaster Management in India

India is a disaster prone country owing to its location in this tropics and in Asia region almost all type disaster natural or manmade are common.

Due to presence of almighty rivers and monsoon rains in the summer's floods are wide spread and almost 12.5% of land of India is flood prone. Its coastline remains under the effect of tropical cyclones throughout the year. Presence of Himalayas in the north and northeast and fault lines responsible for earthquakes about 60% land is vulnerable to earthquake hazard.

Drought prone areas accounts for about 12%.

Being disaster prone region India needs planned disaster management activities.

After tsunami 2004, India decided to establish a separate department of disaster of disaster management as a result NDMA was established in 2005, this step make foundation of a well-organized and well planned disaster management in India, after that budgets were allocated and disaster planning was become an integral part of five years plan.

Table - 11: Some Major Disasters in India

S. No.	*Name of Event*	*Year*	*State & Area*	*Fatalities*
1.	Floods	October 2014	Jammu & Kashmir	-
2.	Cyclone Hud Hud	September 2014	Andhra Pradesh & Odisha	-
3.	Odisha Floods	October 2013	Odisha	21
4.	Andhra Floods	October 2013	Andhra Pradesh	53
5.	Cyclone Phailin	October 2013	Odisha and Andhra Pradesh	23
6.	Floods/Landslides	June 2013	Uttarakhand and Himachal Pradesh	4,094
7.	Cyclone Mahasen	May 2013	Tamil Nadu	08
8.	Cyclone Nilam	October 2012	Tamil Nadu	65
9.	Uttarakhand Floods	Aug – Sep 2012	Uttarkashi, Rudraprayag and Bageshwar	52
10.	Assam Floods	July – Aug 2012	Assam	---
11.	Cyclone Thane	December 2011	Tamil Nadu, Puducherry	47
12.	Sikkim Earthquake	September 2011	Sikkim, West Bengal, Bihar	60
13.	Odisha Floods	September 2011	19 Districts of Odisha	45
14.	Sikkim Earthquake	2011	North Eastern India with epicenter near Nepal Border and Sikkim	97 people died (75 in Sikkim)
15.	Cloudburst	2010	Leh, Ladakh in J&K	257 people died
16.	Drought	2009	252 Districts in 10 States	-----
17.	Krishna Floods	2009	Andhra Pradesh, Karnataka	300 people died
18.	Kosi Floods	2008	North Bihar	527 deaths, 19,323 livestock perished, 2,23,000 houses damaged, 3.3 million persons affected
19.	Cyclone Nisha	2008	Tamil Nadu	204 deaths
20.	Maharashtra Floods	July 2005	Maharashtra State	1094 deaths 167 injured 54 missing

(Contd...)

21.	Kashmir	2005	Mostly Pakistan, Partially Kashmir	1400 deaths in Kashmir (86,000 deaths in total)
22.	Tsunami	2004	Coastline of Tamil Nadu, Kerala, Andhra Pradesh, Pondicherry and Andaman and Nicobar Islands of India	10,749 deaths 5,640 persons missing 2.79 million people affected 11,827 hectares of crops damaged 300,000 fisher folk lost their livelihood
23.	Gujarat Earthquake	2001	Rapar, Bhuj, Bhachau, Anjar, Ahmedabad and Surat in Gujarat State	13,805 deaths 6.3 million people affected
24.	Orissa Super Cyclone	1999	Orissa	Over 10,000 deaths
25.	Cyclone	1996	Andhra Pradesh	1,000 people died, 5,80,000 housed destroyed, Rs. 20.26 billion estimated damage
26.	Latur Earthquake	1993	Latur, Marathwada region of Maharashtra	7,928 people died 30,000 injured
27.	Cyclone	1990	Andhra Pradesh	967 people died, 435,000 acres of land affected
28.	Drought	1987	15 States	300 million people affected
29.	Cyclone	1977	Andhra Pradesh	10,000 deaths hundreds of thousands homeless 40,000 cattle deaths
30.	Drought	1972	Large part of the country	200 million people affected

Surce: NIDM.

National Disaster management authority (NDMA): was formulated under the prime minister of Indian, to create preparedness for disaster and incorporate mitigation measures which were at very poor state till that time, if the time of Tsunami 2004 we were are prepared as we are now then we lose much less than what we lost at that time.

The govt. of India enacted disaster management act in 2005. As per the guide lines of the Act, a National Disaster Management Authority (NDMA) headed by the Prime Minister has been constituted with a vice chairman and five others as members.

The national executive committee (NEC) to be headed by the union home secretory and other secretaries in charge of major departments at the national level and the integrated defense staff will assist the NDMA in its functioning and implementation of the politics and plans of NDMA, and make ensure the compliance of directions issued by the government and NDMA for disaster management in the country.

State Disaster Management Authority (SDMA)

It is a proposal that state level disaster management authority were also constituted in every state headed by the chief minister of the state as chair person and other members to be nominated by them. Now Delhi, Rajasthan, Madhya Pradesh, Assam, Uttra khand, and Nagaland are constituted, other states are also busy in its constitution, in some states like west Bengal and Jammu and Kashmir have home guard civil defense.

State Executive committee (SEC)

SEC will be constituted in every state headed by chief secretory of the state concerned as chairpersons and consisting of concerned secretaries of the states as chairperson and consisting of concerned secretaries to the state governments.

District Disaster Management Authority (DDMA)

DDMA will be constituted in every district in each state under the chairmanship of the collectors and chairpersons of the Zillah Parishad.

Important features of NDMA

1. Prime minister of India is the chairperson of NDMA. Nine members can be nominated by the chairperson, and one vice chairperson.

2. NDMA has responsibility from laying down the guidelines for effective response to disasters, and a holistic and integrated approach will be evolved for disaster management with emphasis on building partnership at various level is required.
3. Community based disaster management including integration of policy plans and execution and capacity development of all spheres.
4. Co-Operation with national and international organizations.
5. Consolidation of pre and post disaster activities.

National policy of Disaster management

Objectives of the policy are as follows:

1. Promotion of preparedness, prevention and resistance at all levels through knowledge, innovation and education.
2. Encouraging mitigation strategies based on technology traditional wisdom and environmental sustainability.
3. Mainstreaming disaster management into developmental planning process.
4. Provision of institutional and technological framework to create regulatory environment and comprehensive regime.
5. Ensuring efficient mechanism for identification, assessment and monitoring of disaster risk.
6. Development of advance forecasting and early warning systems backed by response and safe communication with information technology support.
7. Under taking reconstruction as an opportunity to build disaster resilience structure and habitat for ensuring safer living.
8. Promote the culture of disaster risk prevention and mitigation at all levels promoting a productive and proactive partnership.
9. Strengthen scientific and technical capabilities in all respects.

Functions of NDMA:

1. Lay down policies and approve national plan prepare by ministries and departments.
2. Lay down guidelines for state authority.
3. Provision for disaster management act, co-ordinate the enforcement.
4. Implementation of policy and plan for disaster management
5. The national policy of disaster management concern all the sections of the society including vulnerable group-women, children, elderly, and differently abled persons.
6. Distribution of relief
7. Formation of measures for rehabilitation of affected persons due to disasters.
8. The National disaster policy aims to bring the transparency and accountability in all aspects of disaster management through involvement of community, panchayti raj, local bodies and civil society, community based organizations.
9. India is a disaster prone country owing to its location in this tropics and in Asia region almost all type disaster natural or manmade are common.
10. Due to presence of almighty rivers and monsoon rains in the summer's floods are wide spread and almost 12.5% of land of India is flood prone. Its coastline remains under the effect of tropical cyclones throughout the year. Presence of Himalayas in the north and northeast and fault lines responsible for earthquakes about 60% land is vulnerable to earthquake hazard.
11. Drought prone areas accounts for about 12%.
12. Being disaster prone region India needs planned disaster management activities.

13. After tsunami 2004, India decided to establish a separate department of disaster of disaster management as a result NDMA was established in 2005, this step make foundation of a well-organized and well planned disaster management in India, after that budgets were allocated and disaster planning was become an integral part of five years plan.

The NDRF (National Disaster Response Force):

NDRF is highly professional force developed for the disaster response by the government of India. The force has established itself as a most reliable disaster response force to work in disaster situations. The force is single largest body of its kind in the world no other country in the world has special task force for disaster response like India, as in other countries police and army do this work.

NDRF is acknowledged worldwide for its out standing work in Nepal earthquake of 2015. Apart of it NDRF did a plausible work in the devastating floods of Bihar, UP, and MP in 2016, J & K floods 2014, Chennai Floods of 2015 etc. And disasters like building collapse and train accidents. NDRF has saved more than 5,52,018 lives on various disasters within India and abroad.

Role of institutions in disaster management

Disaster are natural and manmade are inversely at an alarming rate clamming large number of human lives and property losses. Most disaster occur in Asia pacific region. The Indian sub continental is highly vulnerable to floods, droughts, cyclones and earthquakes, landslides etc. Disaster have wide effects on the overall development scenario of the countries. In many parts of the world particularly in Indian sub –continent disasters are main cause of lander development as they push the people into a vicious cycle.

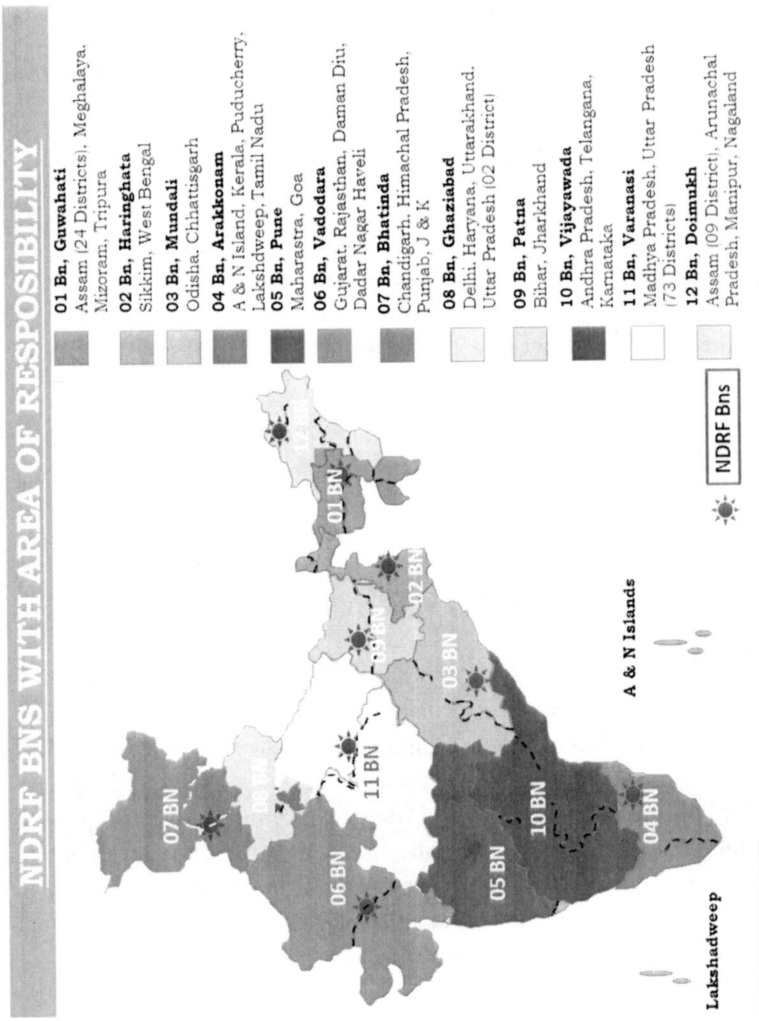
NDRF BNS WITH AREA OF RESPOSIBILITY
01 Bn, Guwahati
Assam (24 Districts), Meghalaya, Mizoram, Tripura
02 Bn, Haringhata
Sikkim, West Bengal
03 Bn, Mundali
Odisha, Chhatisgarh
04 Bn, Arakkonam
A & N Island, Kerala, Puducherry, Lakshdweep, Tamil Nadu
05 Bn, Pune
Maharastra, Goa
06 Bn, Vadodara
Gujarat, Rajasthan, Daman Diu, Dadar Nagar Haveli
07 Bn, Bhatinda
Chandigarh, Himachal Pradesh, Punjab, J & K
08 Bn, Ghaziabad
Delhi, Haryana, Uttarakhand, Uttar Pradesh (02 District)
09 Bn, Patna
Bihar, Jharkhand
10 Bn, Vijayawada
Andhra Pradesh, Telangana, Karnataka
11 Bn, Varanasi
Madhya Pradesh, Uttar Pradesh (73 Districts)
12 Bn, Doimukh
Assam (09 District), Arunachal Pradesh, Manipur, Nagaland
NDRF Bns
07 BN
06 BN
05 BN
10 BN
04 BN
11 BN
03 BN
02 BN
01 BN
12 BN
A & N Islands
Lakshadweep

In the decade between the 1990-2000, the United Nations, International Decade for natural Disaster Reduction (IDNDR) was observed globally, during this period through Yokohama strategy stress was on a holistic approach the disaster preparedness and mitigation measures against the disaster.

Education institution faculty and students can play a very important role in disaster management which can be a channel for awareness generation in the society. The most organized sectors of sectors of students i.e. NCC and NSS can play an effective role for the purpose. They may be incorporated into the local land relief and awareness programs.

Educational institutions can contribute in disaster management through three ways:

1. Research: Institution are busy in various types of studies which can be form a basis for public policy decisions.
2. Technical Advice: Faculty from IIT and the institution are often tapped issues related to disaster management issues. Their opinions are respected and valued by the people at large.

Through education and training knowledge is transferred to professionals from various fields including engineering, medicine, economics public administration and polity.

In India IIT Rourkee, the Anna University, Jamia Millia Islamia, university Delhi and IGNOU are some important institutions which are working on issues related to disaster management and climate sustainability.

Advantages of improvement of educational institutions in disaster education.

Following advantages can be seen:

1. It provides relevant information about the environment and disasters prone to a particular region.

2. It prepares students for disaster response and recovery activities.
3. They use past experiences for describing risks and vulnerabilities and suggest preventive measures to combat disasters.
4. It helps to develop a spirit of collective work as successful disaster manager.
5. It promotes decision making during a disaster situation.

   In some colleges UGC has initiated the summer school program for teachers of colleges.

   In India important departments engaged in disaster management are National Institute of Disaster Management, NIDM, Natural Disaster Management (NDM), State institute of Rural development (SIRDS), Central Road Research Institute (CRRI), India Meteorological Department (IMD), Central Water Commission(CWC), IITs etc.

Strengthening the institutional mechanism

For disaster Preparedness and mitigation institutional strengthening is of vital importance. Government of India established national Centre for disaster management in Indian institute of public administration, New Delhi, now it is known as National Institute of Disaster Management, NIDM since October 16, 2003. This institute is busy in human resource development, research and training, and preparation of data base with the help of government and NGOs sectors. Similar centers are established in various states steps are also taken to involve IITs, universities, research and development and leading NGOs, for supplementing the government efforts in disaster management.

Role of Panchayats in Disaster Preparedness

For disaster planning local bodies like panchayats have not been consulted in preparedness, relief and rehabilitation work. The local bodies like panchayats should be encouraged and empowered to manage local affairs with the available resources. The elected leaders of panchayats can be trained to handle crisis in every phase of disaster management.

The role of panchayats is important in view of their proximity to local community. It is possible to ensure accountability and transparency through institution of gram Sabah. In the event of a disaster all people are affected in that zone but poor and vulnerable people suffer most. The benefits of community participation are:

(a) People can become sources of useful ideas, particularly indigenous technical knowledge and skills.

(b) New knowledge, skills and techniques used by a group can set an example for others

(c) Community can participate in decision making process with regards to development schemes for disaster reduction measures. In this way people‘s choices can approximately be reflected in development programs.

(d) Disaster leaders need to be trained to evolve a community based disaster preparedness and mitigation plan.

(e) With social mobilization process, efforts should be made to find out how the local institutions be effective involve to complement the modern practices in the disaster mitigation efforts.

(f) It can develop community based monitoring system to ensure effective delivery of disaster mitigation strategies at local level.

Disaster Preparedness

Disaster preparedness is an effective way to reduce the impact of disasters, it acts as a link between emergency responses. preparedness is a pre disaster phase in which we try to prepare our self for disaster, it involves forecasting and taking precautionary measures, prior to be threat when advance warning are possible preparedness also incomes the regular testing of warning systems and plans for evaluation and other measures to be taken to minimize potential loss of life and physical damage. It also involves the education and training of personnel's and population and population at risk.

Capacity development covers strengthening of institutions mechanism and capacities stakeholders at all levels. Strengthening of the national, regional and nodal capacity-building institution and resource centers at distinct and state level.

Education and training includes inclusion of disaster management in the curriculums of universities to make aware people about how to response disaster. Mapping of vulnerable groups long terms risk reduction measures are taken to minimize the adverse effects, of a hazard by eliminating the vulnerabilities and losses faced by a community. The main object of disaster preparedness is to make certain group of people or property safe and less vulnerable for the impact of hazards. Disaster preparedness must be an active and on-going process.

Preparedness plans must be dynamic velarizes, which need timely review, modification, updating and testing by experts.

Passive disaster preparedness includes preparation of disaster manuals, stockpiling of relief goals, development resource inventory active disaster preparedness includes comprehensive response plans, monitoring hazard threats, education and training of personnel and peoples of community at risk.

Preparedness activities in pre disaster period

- Formulation of plans at national, state and district and community levels.
- Awareness generation and dissemination of information through media and other means about the potential hazards and their associated risks.
- Installation of appropriate forecasting and warning system.
- Strengthening of physical infrastructure.
- Evacuation of people in safer areas during the emergency.

Preparedness during the emergency:

- Provision of food, shelter and provision of medical facilities.
- Security arrangements for the displaced people.

During the post disaster period:

- Rescue operations for the victims.
- Restoration of communication system.
- Reconstruction and rehabilitation work.

Disaster preparedness measures

Disaster Mapping

Is done to assess the impact of disaster on population, property and natural resources, disaster mapping also helps in risk assessment, rescue and relief operatNGOs can play an important role in ions.

Disaster Preparedness Plan

Planning make any operation organized and more target oriented both short term and long term plans are made to encompass the aims and objectives, the organizational structure for tackling disasters, preparedness measures, the communication system, warning arrangements, community disaster operations. Planning in the context of disaster preparedness is holistic. An effective plan gives a clear and coherent approach for dealing with disasters.

The purpose of planning is to anticipate future situations and requirements, thus ensuring the application of effective and coordinated counter-measures.

NGOs Participation in Disaster Preparedness Plan

NGOs can play an important role in disaster preparedness if specific tasks are given to them, NGOs should work with dedication and without self-interest.

Disaster Preparedness for people with special Needs and for vulnerable groups

Globally it is observed that it is observed that there is ascertain section of the society who are more susceptible. These vulnerable sections lack resources and are deprived of basic necessities. These include women especially the pregnant, lactating, children, old people, physically challenged, land less and poor farmers.

Incident Command System

In order to professionalize the response system, it is to propose the incident command system in country the system provides with most experienced and knowledgeable persons.

Remote sensing and GIS database is an effective tool for emergency responds, to assess the information in terms of crucial parameters are location of public facilities, communication links, and disaster sites.

Disaster preparedness focus on warning and forecasting of disaster and warn people in advance about the occurrence of disaster particularly rapid onset disasters. The space technology and the disaster preparedness community can work together to develop an effective and accurate, method for mitigation of disasters. Disaster prevention is a long term phenomenon which can be best studied with the help of satellite monitoring of fault lines, flood plains, landslides, land use changes etc.

Satellite monitoring involves assessment of damage occur during a disaster.

Vulnerability index of India

India is a vast country having 28 states and seven Union territories with 593 districts. India's landscape is highly vulnerable to disasters because of both geophysical as well as socio-economic factors, disadvantaged people are more vulnerable to disaster such as poor, illiterate, females, children, and handicapped, aged. About 45 % of the total population of India is living below poverty line, having less than an income of Rs. 20 per day per capita. Poverty accelerate the rate of disasters.

Natural Disaster in India

Table 14: Population killed and affected in Disaster in India (1993-2015

Total population killed (1993-2005)	*Total population affected (1993-2005)*	*Total population killed (2006-2015*	*Total population affected (2006-2015)*	*Total population killed (2015)*	*Total population affected (2015)*
50,846	315,125,127	64478	659134721	2044	8,997490

Source: World Disaster Report 2016.

The above figures are enough to show the high vulnerability of India for disasters.

India has a high rank in the world records on ground of many disasters like cyclones earthquakes and floods. India among top ten world disaster countries in case of cyclones, avalanches and landslides. Many deadly cyclones killed lakhs of people in the past invaded India. Snowstorm and avalanche are active disasters in the snowbound areas caused many deaths annually. In the monsoon season in Himalayan areas landslides and mud slides caused many causalities and injuries. In the Landslides and Avalanche related deaths India ranked second and ranked seventh in the fatalities of cyclones. India Ranked at 77th position on the world risk index, according to world disaster report 2016.

Table 15: India vs world. Number of people killed and affected by (1993-2015)

Continent/ Countries	*Population Killed (1993-2005)*	*Population Affected (1993-2005)*	*Population Killed (2006-2015*	*Population Affected (2006-2015)*	*Total population killed (2015)*	*Total population affected (2015)*
India	85732	713,342,088	26,895	132,164,431	3666	16561182
Asia	1043,475	2416,805,041	390,054	1523,331,637	21770	68,083,036
China	33,295	1,285,225,236	111,846	1,019,008,563 1	1929	3,779,096
United States of America	6153	10463765	3538	15099494	335	49699
World	**1281837**	**2802,160,490**	**777191**	**1917,556,951**	**32550**	**108,493,144**

Source: World disaster Report-2016.

India vs rest of World number of people killed and affected

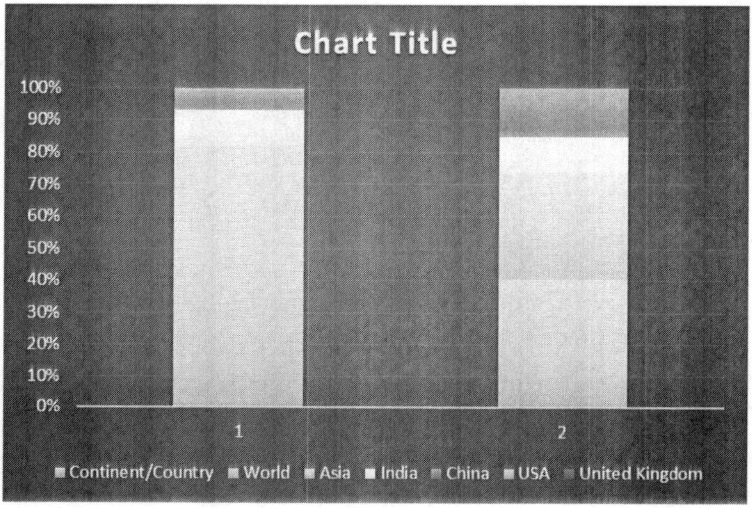

Source: Created by the author.

Fig. 24: The figure shows number of people killed and affected in natural disasters in India As compare to the world.

Earthquake

In India about 60% land is prone to earthquakes, in last two decades India witnessed some devastating earthquakes such as earthquakes of Latur, Bhuj, Uttrakhand, Kashmir etc.The Bhuj earthquake on 26 January 2001 raised the questions on our preparedness to respond to disasters.

Himalayas lying in the northern part of India are the young fold mountains, therefore geologically very active. The peninsular part is crystalline it is a stationary zone, considered seismically least active.

Seismic Zonation of India

Seismic zonation is a method of classification of areas on the basis of seismic activity in the areas. Seismic zonation can be done at national, regional or local level.

A seismic zonation map specifies the level of force or ground motion for earthquake resistance design. The seismic zoning is a multidisciplinary task which involves, geologists, geographers, seismologists, geophysicist and structural engineers.

Need of seismic zonation

1. These maps are based on geology and past experience s of the area so they give a description of type of risk.
2. It helps in land use planning and gives a comprehensive information for planning and development.
3. Maps of probabilistic hazards gives an idea of underlying statistical uncertainty, these maps gives odds at which specified earthquake intensity would be exceeded at a site of the interest within a given time period.
4. Seismic zonation is used to reduce the human and economical losses caused by earthquake and in this way the economic development and political stability of vulnerable areas can be

Types of zonation

The seismic zonation can be done on the basis of the following scales:-

1. Macro zonation
2. Micro zonation
3. Nano zonation

(i) Macro zonation:- Earthquakes and related risk are mapped at sub-regional level, as geoscientists are interested in minute details they work at very small scale covering large areas and development planners work at large scales covering small areas. The scale for macro zonation is generally used as 1:10,000,00.

(ii) Micro zonation:-Micro zonation use 3 scales they are:-

Level I: 1:10,000,00 to 1:50,000, Level II: 1:10,000,00 to 1:10,000.

Level III: which assist the complete zoning activities 1:25,000 to 1:5,000.

(iii) Nano zonation:- It is a detailed zonation based on argons methods the scale generally taken is less than 500. This level of zonation requiring detailed site specific information it involves high value of money but if hazard potential is very high and the proposed development is regarded as critical, high level of investment can be afforded.

Many large earthquakes occurred in India.

Earthquake engineering in India was developed so early earthquake resistant buildings were started to build after 1897 Assam earthquake, which is still prevalent in the north east India. First seismic hazard map was prepared in 1935 by biological survey of India.

In 1962 BIS published the seismic zonation map of India, Based on earthquake epicenters and a seasonal map published by GSI in 1935.

The first Indian seismic code was published in 1962, which provided criteria for building bridges, liquids retaining for tanks, gravity tanks etc how to make earthquake resistant. In this map the Himalayas and North East India were designated one vii-iv and Deccan plateau was marked as zero.

The first zonal map divided country into 7 seismic zones 0-6 corresponding to areas liable to maximum intensely.

In 1967 another codes were published and reversed in 1976 and 1993. It outlines the seismic designs and construction requirements. A comprehensive manual was published in 1983.

In 1976 after koyana earthquake intensely necessity arose to revived the criteria of zonation particularly in the Deccan plateau, it reduces the number of zone from 7 to 5.

Zone v: is the maximum intensely zone comprises entire northeast India parts of Jammu and Kashmir, Himachal Pradesh, Uttaranchal, Rann of Kutch in Gujarat, parts of north Bihar and Andaman and Nicobar Island.

Zone IV: Delhi, Sikkim, northern parts of U.P, Bihar and west Bengal, parts of Gujarat.

Zone III: comprises of Kerala, Goa, Lakshadweep, Madhya Pradesh, Jharkhand, Chhattisgarh, Maharashtra, Orissa, Andhra Pradesh, Tamil Nadu and Karnataka.

Zone II: covers all the remaining parts of the country.

According to the government analysis and seismic revision, major cities under earthquake threat are:

Zone V cities – Guwahati and Srinagar

Zone IV cities_ Delhi, Patna, Dehradun, Jamnagar, Meerut, Jammu, Amritsar and Jalandhar

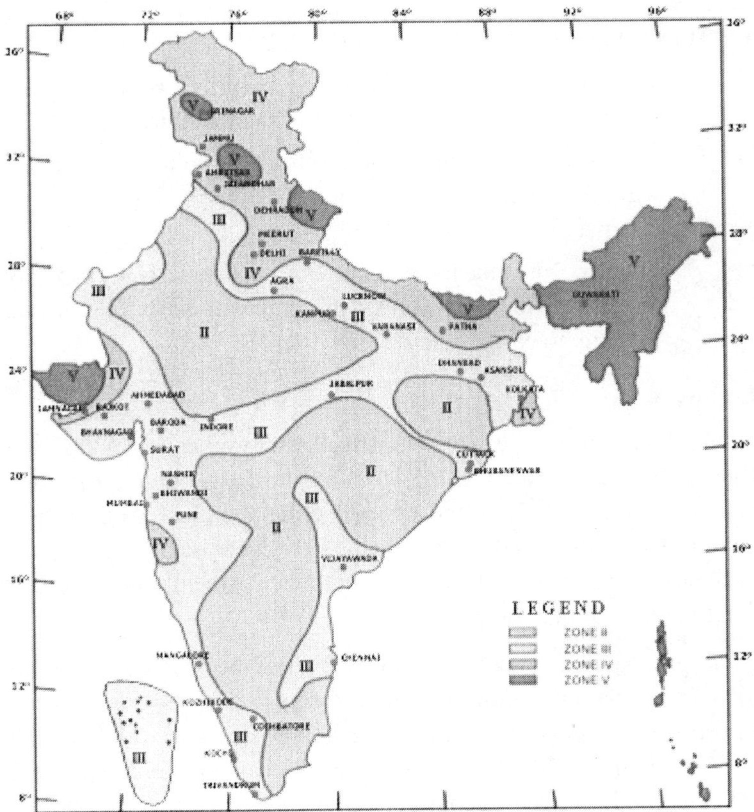

Bureau of Indian Standards Criteria for earthquake resistant design of structures IS 1893 : 2002

The above map shows seismic zones of India based on bureau of Indian standards Criteria.

Preparedness for the threat is the only solution to reduce the impacts as this is beyond human control.

In 1935, Kangra was hit by a magnitude 8.6 quake that killed at least 19000 people. An earthquake hit Bihar in 1934 at magnitude 8.4 shook buildings from Mumbai to Lhasa. The destruction in Bihar was great 4, 000 people were killed.

Latur was hit in 1993 around 10000 people were killed.

Bhuj earthquake in 2001 killed at least 13, 800 people.

Factors affecting the structural damage

- The intensity of earth shaking is indicated by the ground accelerations caused at the base of the structure.
- The dynamic parameters of the structure such as the mass of various elements, their stiffness, and deformation and energy dissipating capacity.
- The strength of the foundation soils, the load resisting capacity of the individual elements, their connections and the whole capacity to carry the earthquake forces.

Lightening

Lightning kills more people in India than any other natural disaster. Every year, 1,755 people in the country die from lightning strikes, according to an article in the Economic and Political Weekly. In a country where thousands of people die every year because of floods, cyclones, landslides and other such events, the death toll from lightning is far more present a sad picture.

Lightning is responsible for at least 10% of the total deaths caused by natural calamities in India and, according to the National Crime Records Bureau data, at least 2,000 deaths are associated with lightning every year since 2005. Still, lightning is not categorized as a natural calamity, and affected people or their families are not eligible for compensation from the government unlike in the case of floods or earthquakes.

The situation is particularly bad in the eastern states of Assam, West Bengal and Odisha and Bihar, where the highest number of people died every year because of lightening. Many examples are there among which important one are on 27 May 2017, the Lightning and thunderstorms have killed 29 people and injured many others in Indian state of Bihar, according to disaster management officials.

Mostly deaths took place in village areas across the districts of Jamui, Champaran, Vaishali and Samastipur.

In similar dangerous storms in April 2015, 56 people died because of lightening. Storm and winds in Bihar are commonly blowing from April 20 to June 30 until the monsoon starts. Source Al Jazeera and news agencies.

Three people were killed and at least 50 injured when lightning struck Assam's Cachar district on Wednesday. Most of the victims are farm labourers who take shelter under isolated trees during a thunderstorm. Northeastern states, Maharashtra, Kerala, Jharkhand and Bihar also suffer heavy casualties.

Risk Reduction

To reduce the fatalities from lightening the risk could be reduced by preventive measures.

Warnings are issued by the Indian Meteorological Department regarding storms and lightning strikes. Unfortunately India does not have a lightning detection network. Like other developed countries of America, Canada and Europe.

To reduce the losses every state government can identify *vulnerable* areas in their states this can be done on the basis of geographical aspect of the areas. These could be mapped for any natural hazards.

The government could also issue guidelines on what to do when there is lightning.

Now the state governments are issuing advisory regarding the hazard, the Bihar government, for the first time, issued advisory in the media about how residents should deal with a thunderstorm.

Public awareness programmes are a good source of information for the training of people how to deal with the problem, street plays, written material and media can play an important role in disseminating information to the community.

Doe's and Don'ts during a thunder storm.

1. Go indoor to find safe shelter
2. Crouch down in a ball like position
3. Do not lie down.
4. If in a group stay from each other.
5. Get away from open areas.
6. Avoid being at high elevations like mountain top.

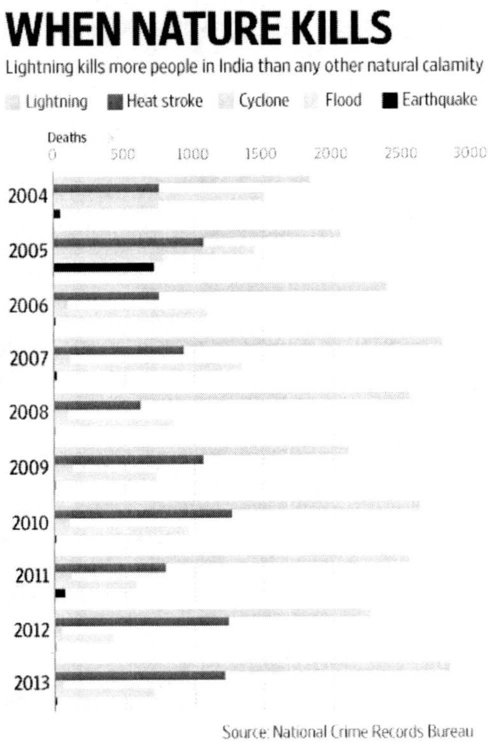

Fig. 25: India verses World Natural disasters.

7. Never use plugged in phones.
8. Avoid standing under trees or tall buildings.

The incidences and deaths are distributed, total impact of lightning is more than any natural disaster. Attention needs to be paid to this issue.

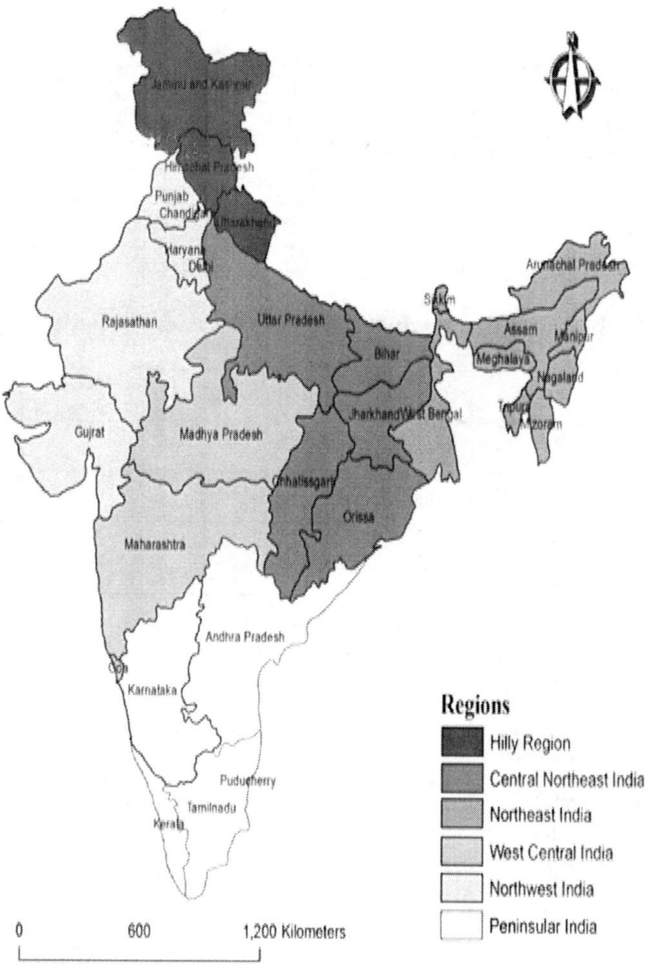

Source: based on the data provided by IMD.

People Killed by Lighetning

Date and year	*Place and state*	*Number of victims*
11 August 2010	Sapaul (Bihar)	52
18 July 1996	Hazaribagh (Jharkhand)	39
28 September 2004	Nashik (Maharashtra)	31
25 June 2006	Gwalior (Madhya Pradesh)	30
13 March 2007	Balia and Basti (Uttar Pradesh)	28
6 July 2000	Balia (Uttar Pradesh)	27
6 June 2004	Basti and Gonda (Uttar Pradesh)	25
30 June 2007	Raigad (Maharashtra)	25
19 June 2006	Hingoli (Maharashtra)	24
27 June 2010	Yavatmal (Maharashtra)	24

Table 2. Top 10 fatality incidents recorded in India during the period 1979–2011

Source: Disastrous Weather Events, India Meteorological Department, Pune.

Table 16: Damages due to Natural Disasters in India.

Date	*Location*	*Mag.*	*I*	*Deaths*	*Injuries*	*Total damage / notes*	
2017-01-03	India, Bangladesh	5.7 M_w	V	3	8		
2016-01-04	India, Myanmar, Bangladesh	6.7 M_w	VII	11	200		
2015-10-26	Afghanistan, India, Pakistan	7.7 M_w	VII	399	2,536		
2015-05-12	Nepal, India	7.3 M_w	VIII	218	3,500+		
2015-04-25	Nepal, India	7.8 M_w	IX	8,964	21,952	$10 billion	
2013-05-01	Kashmir	5.7 M_w		3	90	$19.5 million	NGDC
2011-09-18	Gangtok, Sikkim	6.9 M_w	VII	>111			
2009-08-10	Andaman Islands	7.5 M_w	VIII			Tsunami warning issued	
2008-02-06	West Bengal	4.3 M_b		1	50	Buildings damaged	NGDC
2007-11-06	Gujarat	5.1 M_w	V	1	5	Buildings damaged	[2]
2006-03-07	Gujarat	5.5 M_w	VI		7	Buildings damaged	[3]
2006-02-14	Sikkim	5.3 M_w	V	2	2	Landslide	[4]
2005-12-14	Uttarakhand	5.1 M_w	VI	1	3	Building destroyed	[5]

2005-10-08	Kashmir	7.6 M_w	VIII	86,000–87,351	69,000–75,266	2.8 million displaced	
2005-03-15	Maharashtra	4.9 M_w	VII		45	Buildings damaged	[6]
2004-12-26	off northern Sumatra	9.1–9.3 M_w	IX	230,000–280,000		Destructive tsunami	
2002-09-13	Andaman Islands	6.5 M_w		2		Destructive tsunami	NGDC
2001-01-26	Gujarat	7.7 M_w	X	13,805–20,023	~166,800	Republic Day (India)	
1999-03-29	Chamoli district-Uttarakhand	6.8 M_w	VIII	~103			
1997-11-21	Bangladesh, India	6.1 M_w		23	200		
1997-05-22	Jabalpur, Madhya Pradesh	5.8 M_w	VIII	38–56	1,000–1,500	\$37–143 million	
1993-09-30	Latur, Maharashtra	6.2 M_w	VIII	9,748	30,000		
1991-10-20	Uttarkashi, Uttarakhand	6.8 M_w	IX	768–2,000	1,383–1,800		

Source: Wikipedia.

India's rank is 77th on the World Risk Index, (topped by Island state of Vanuatu). The World Risk Report 2016, published by the UNU Institute for Environment and Human Security (UNU-EHS) and Bundnis Entwicklung Hilft, this analysis is mainly based on the quality of infrastructure each country have contribute to a country's disaster risk.

The calculated Index ranks 171 countries according to their risk levels based on the strength of their infrastructural facilities available to face different natural hazards such as floods, cyclones, or earthquakes.

The report said that inadequate infrastructure and weak logistic management increase the risk manifold that an extreme natural event become a disaster.

Floods

Flood is temporary inundation of water for some days. India is a worst flood affected country in the world after Bangladesh and accounts of one-fifth of the global death count due to floods.

During summer season maximum rainfall occur July-September as a result Rivers receive heavy discharge which leads to widespread flooding. Northern plains especially north –eastern plains are highly prone to annual flooding. The flooding is aggravated by sedimentation of channels, drainage congestion and synchronization of river floods. The upper reaches, Ganga and Brahmaputra basin are the most flood prone areas of Indo- Himalayan Rivers carry a lot of sediments and rode their banks in the Gangetic plain.

Fig. 26: Flood in Uttrakhand.

Other areas are north-west region where Tapi and Narmada rivers flows, the central India and Deccan with the rivers Mahanadi, Krishna and Kaveri. Annually about 8 million hectares and approximately 3.7 million hectares average cropped area affected annually.

Most flood prone areas of India are West Bengal, Orissa, Andhra Pradesh, Kerala, Assam, Bihar, Gujarat, Uttar Pradesh, Haryana and Punjab.

Physical Causes of floods

- Low-lying flat topography in the flat terrain rivers have a tendency to meander in the flood plain many rivers flood due to this reason for example River Kosi in Bihar and Brahmaputra in Assam.
- Excessive rainfall during monsoon season- 75% of the total rainfall occur in July in monsoon season causes heavy discharge of water in the rainy season.
- Heavy snow melting in Himalayas.
- Landslide in the river basins.
- Choking of channels and siltation of river beds.- caused by physical hazards earthquakes and landslides as result rivers flood. Due to steep gradient in the Himalayan region rivers erode huge amount of material and bringing down with them, as the velocity of water reduces they deposit it in their beds which reduces the water carrying capacity of the rivers.
- Tidal and wind effects-Sea tides deposit silt on the river mouths, obstructions like embankments, railways etc. case obstruction in free flow of rivers.
- Deforestation causes increase in the frequency of water flow and decrease in the infiltration, deforestation of hill slopes lead to speedy runoff which raised the level of water.

The unchecked and haphazard development of flood plains, can be taken as the main factor is the cause of ever increasing flood damage in various parts of the country. After investing huge investments in the flood risk reduction in the past but no flood control structure can be considered to provide total or absolute protection against all magnitudes of floods, the costs are far more than benefits. Over the last two decades the central India has facing weather related problems like torrential rainfall and the flash floods, the most affected flood areas of India are river banks and deltas of Yamuna Sutlej, Ganga, Ghaghara, Kosi, Brahmaputra, Ravi, Mahanadi, Damodar, Godavari, etc.

Table 17: State wise flood prone Areas.

State/Union Territory	***Flood Prone***
Uttar Pradesh	7.3
Bihar	4.2
Punjab	3.7
Rajasthan	3.2
Assam	3.1
West Bengal	2.6
Haryana	2.3
Orissa	1.4
Andhra Pradesh	1.4
Gujarat	1.4
Kerala	0.8
Tamil Nadu	0.4
Tripura	0.3
Madhya Pradesh	0.2
Himachal Pradesh	0.2
Maharashtra	0.08
Himachal Pradesh	0.08
Jammu & Kashmir	0.08
Manipur	0.05
Delhi	0.02
Karnataka	0.02
Meghalaya	0.01
Pondicherry Total	**0.01 33.5**

Highest flood prone regions of India:

There are three important areas:

(1) Ganga Basin: Drained by Ganga and its tributaries, affected states are West Bengal, Bihar, and Uttar Pradesh. Besides the Ganga other rivers like Sharda, Rabti, Ghagra and Gandak are the cause of heavy flooding in eastern Uttar Pradesh. The Yamuna is famous for floods in Haryana and Delhi. Bihar faces massive destruction every year due to Kosi River. Burhi Ganga, Baghmati etc. floods in West Bengal.

(2) Brahmaputra and Barak basin: This basin is drained by these two rivers in north eastern states like West-Bengal,

Assam and Sikkim. Tieststa, Tasha in north and west Bengal and Manipur.

(3) Central India and Deccan River Basin: In this region River Mahanadi, Brahmini cause havoc. The delta of these three rivers is thickly populated, even some small rivers of Kerala and mud streams from nearby hills.

Southern and central India observe floods caused by Narmada, Tapti, Krishna, Godavari, Mahanadi, Krishna causes floods in the coastal region of Andhra, Orissa, Tamil Nadu.

Table 18: The Worst Floods In India

Year	*Deaths*	*Total Damage (in US dollars)*
1977	11, 316	1202
1988	4252	4630
1979	3637	614
1968	3497	211
1978	3396	1455
2007	3389	13, 393
1998	2889	8861
2008	2876	9566
1993	2864	3283
2000	2606	8865

Source: Centre water commission.Notwithstanding flood policy and flood prone areas. The distribution of damage is wide spread.

Cyclones

Cyclone is a meteorological phenomenon, intense low pressure in the centre rotating in cyclic manner, originate over the open oceans and move towards the land. They invade shorelines first then moves into the interior of the land.

North West pacific region is the most cyclone affected area in the world.

The eastern coast of India is more vulnerable to cyclone disasters than west, number of cyclones originate in Bay of Bengal is more than the Arabian Sea and also they are more intense than Arabian Sea cyclones. India experience cyclones generally in April-May and October – November.

India has 7,516 km long coastline exposed to approximately 10% the tropical cyclones of the world, of these mostly have their initial genesis over the Bay of Bengal and strike the eastern coast of India.

Source: NDMA

On an estimate around 5-6 cyclones occurred every year of which 2-3 could be of severe intensity cause great losses. The ratio of occurrence of cyclones in the Bay of Bengal and Arabian Sea is 4:1.

A study shows that between 1891-2015 nearly 4000 cyclones were occur of which 100 were severe in nature and occur in 50 kms wide strip on the east coast, on the west coast only 40 cyclones were occurred in the same time period out of which only 22 were severe.

They are proved more disastrous at the time of rainfall in north Indian Ocean, as they are accompanied by strong winds, storm surges and torrential rainfall.

Storm surges created by cyclones are the greatest killer as through it water enters in the coastal areas, and causes heavy flooding, beaches vegetation and embankments are destroyed and soil erosion is increases.

Areas prone to be tsunami and cyclones

Cyclone Warnings

Cyclones can be tackled from their point of origin by satellites, meteorological satellites recording data round the clock, these satellite imageries play important role for monitoring of storms, strength and intensity of the storm is also calculated due to which estimate of losses can be calculated in advance. The infrared polar orbiting satellites observe day and night and geostationary satellite afford continuous coverage during day time. A network of 10 cyclone detection radars covering entire coast and west coasts is being used for cyclone warning, their movement is monitored by IRS and NOAA satellites, warnings issued by warning centres located at ACWS, located at Kolkata, Madras, Mumbai, Bhubaneswar, Vishakhapatnam and Ahmadabad around 100 disasters warning systems have been installed in cyclone prone villages of Andhra and Tamil Nadu, Orissa and West Bengal.

Management of cyclone hazards

The satellites data provide information regarding flood extent, destructed villages, estimation of loss of crops, and satellite can also use in providing restorative relief measures.

Mapping of cyclone affected areas, damage assessment of property, communication facilities and agricultural production can be easily done with the help of satellite data.

The operational use and the commissioning of DWS in 1987 has opened the door for disaster mitigation mechanism. (Rao 1996) DWS system was successful used on Andhra coast on May 9, 1990 when 1, 70000 people were evacuated, and in this way thousands of life were saved.

Meteorological satellite data are extremely useful for cyclone warnings and monitoring.

An advantage of earth observation satellite data has been demonstrated during Orissa Super cyclone event, when a storm hit the Para Deep (Orissa) on October 1999 caused heavy loss of life and property.

The National remote sensing agency acted immediately and inundated areas were mapped, pre cyclone IRF, LISS- III data collected, on October 1999, such map was used by Orissa government for relief operation later on, on the basis of this data damage assessment was also done and provided to the Orissa government.

In a recent example cyclone phalian was going to hit Andhra coast, one of the country's largest cyclone was tracked and warnings were issued and 440, 000 people were evacuated from the vulnerable areas. Cyclone phailian left a trail of destruction along the east coast of India, but advanced warnings and evacuation saved hundreds of thousands of lives. India evacuated more than half a million people, it was the biggest evacuation in India's history.

Drought

Drought occur when aridity increases over humidity due to scarcity of water. India's 1/6 land area with 12% population is drought prone with annual rainfall up to 60 centimeters, the irrigation commission in 1972 had identified 67 districts as drought prone. They are located in eight states covering an area of 49.7 million hectares.

Later based on detailed studies 74 districts of the country have been identified as drought prone. These districts cover arid and semi-arid areas cover an area of 0.6 million sq.km First region comprised semi desert and desert areas, in this region rainfall is less than750mm or 75 cm at some places it is even less than 44 cm.

The second tract comprises the regions of eastern part of Western Ghats, 300km away from the coast known as the rainbow area of the Western Ghats. Rainfall is less than 75cm and is highly erratic, besides these two regions there are other pockets also of drought prone areas in India. Trinuvali district in Tamil Nadu, Coimbatore area and swarashtra and kachh, Kalahandi region of Odisha all these scattered areas occupied an area of 0.1 million sq.km.

In India monsoon rain is the main source of rainfall, the average annual rainfall in India is 120cms and monsoon contributing about 90cms. The manifestation of monsoons are many such as sometime monsoon delayed, it has long breaks, the distribution of rainfall in India is highly uneven, sometime monsoon doesn't yield sufficient rainfall due to various causes such as El-Nino, southern oscillations and temperature and pressure conditions of Indian Ocean. Any delay or failure of monsoon is responsible for occurrence of drought conditions in India. The worst droughts of last 30 years are those which occur in 1990, 1994, 1997-99 and 2001. This drought was netted in 356 districts of India. The Deccan plateau accounts for 58% drought recurrence. Rajasthan is most drought prone state of India.

The eastern Maharashtra plateau and Karnataka plateau also are the region of severe droughts. Except river deltas and mountain region drought covers all the rest of India mostly in the western parts where amount of rainfall decreases as compare to east. There are two core areas of drought one is the western borders of India which she shares with Pakistan, this is an extremely arid tract includes Rajasthan the Luni river valley and the Gujarat plains. Another core region is the heart of the Deccan from Maharashtra plateau to Karnataka plateau. This is the rain shadow area that runs along the Western Ghats.

Approximately 50 million people are affected by drought every year 6% of country's area is drought prone. The droughts of most severe intensity occur once is 32years and almost every third year is a drought year. In last 100 years Rajasthan faced worst drought in 2002 when monsoon rainfall was 65% below normal, lowest in the last 100 years crop was totally vanished under groundwater table dipped to 3.2-6m below, most of surface water resources were dried, 44.8 million people and 45.20 million livestock were affected. A common policy is evolved to dovetail short-term relief measures into long-term interventions being handled in different ministries and departments for comprehensive drought Management. Global and National best practices in Drought Management are identified and adopted. An active and continuous participation of various States in the country as well as the ministries and departments of Central Government is required for Drought Management.

Objectives of the National Guidelines on Drought Management

Drought management is the prime responsibility of the state governments where it occur. The criteria for drought declaration vary among the states widely. The indicators and methodology for drought intensity assessment and monitoring of droughts also differ from one state to another.

Slow setting in of drought makes it difficult to determine the onset and end of drought.

Drought Management in India

In India, India Meteorological Department (IMD), National Remote sensing Agency (NRSA), Central Water Commission (CWC), and Ministry of Agriculture are responsible for drought monitoring in the country.

The Desert Development Programme: Initiated in 1977-78, covers both the hot desert regions of Gujarat, Rajasthan, and Haryana and the cold desert areas in Jammu and Kashmir

Early Warnings

Early Warnings for drought is not possible only we can assess it on the basis of nature and pattern of precipitation that drought may occur.

We can provide relief to drought affected people, long term measure involves afforestation to enhance the moisture content into the air. Water harvesting to restore the water resources, new verities of crops which will be drought resistant, new agricultural practices to save water. Watershed development. Most notable programs are the National watershed development program and the integrated watershed development program (Mahalawat and Singh, 2003, Singh and Mahalawat 2001, Singh et.al, 2007.

Anthropogenic Disasters in India

Bhopal Gas tragedy

This is the worst ever chemical disaster in the history occurred in Bhopal, India on 3 December 1984 at the union carbide India Limited Pesticide Plant, when methyl Isocyanides (MIC) leaked accidently from a chemical plant, and caused death of over 2500 people by the exposure of poisonous gas while 558125 were injured approximately 3900 severely and permanently disabled.

Methyl isocyanides boils at about 40C at atmospheric pressure. According to reports, the content of the storage tank over heated and boiled, hence the valves are opened, the discharge of vapour about 25 tones, was very large for the capacity of the rubbing system. The escaped gas vapour spread beyond the plant boundary and as a consequence 2500 people died and many thousand were injured by the poisoning. The c overheating was occur due to contamination of the methyl isocyanides by water or some other materials.

The site was closed in 1986. Following the Environment Protection Act passed in 1986, by the Ministry of Environment and Forest which strengthened India's commitment towards environmental protection. In Dec, 2008 Madhya Pradesh High Court decided that the waste should be destroyed at Ankleshwar in Gujarat. There was strong protest across the India by activists and environmentalists. Finally, on 8th June 2012, Incineration of Toxic Bhopal Waste agreed to pay $4.1 million to dispose of UCIL waste in Germany.

Fig 27: Bhopal Gas disaster Source:- Association for Diplomatic Studies & Training

Oleum Gas Leak

Just after Bhopal Gas Leak, M.C. Mehta V Union of India case sprang up. The leakage happened in the heart of capital city. It was due to the aftermath of Oleum gas leak from Shriram Food and Fertilizers Ltd. Complex at Delhi. It created a lot of panic in Delhi when people were only recovering from the shock of Bhopal Disaster only two days before. Shriram Food and Fertilizers was set up in a single complex spreading in about 76 acres. Within a radius of 3kms it was surrounded by colonies such as Punjabi Bagh, West Patel Nagar, Ashok Vihar, Karampura, Shastri Nagar and Tri nagar. One person died and several else were injured in this mishap.

Mumbai Flash floods 2005

Heavy torrential rainfall occur on 26th July when Mumbai came to a 'standstill'. At least 5, 000 people died in the *Maharashtra floods of 2005*. It occurred just one month after the June 2005 Gujarat floods. The floods were caused due to the eighth day heaviest rainfall ever recorded in 24-hours continued for the next day also. One of the key factors aggravating the disaster was Antiquated Drainage System. The present storm-drainage water system in Mumbai was only capable of carrying 25 mm whereas the city received 993mm of total rainfall. Unlike South Mumbai, buildings in Northern Suburb were haphazard and without planning. Mangrove system which existed along Mithi River was the main cause of flooding and Mahim Creek were being destroyed and replaced with construction. Environmentalist and experts claims it to be poor preparedness of government authorities. A lot of illegal construction was done in lower lying areas of Mumbai. Authorities were not trained to deal with such situation. Drainage system was old and blocked with human garbage which caused the mishap.

2010 Ladakh Floods

In 2010 Flash floods occur in Ladakh, because of change in rainfall pattern due to global warming. At least 255 died after a cloudburst and heavy rains which continued the overnight has triggered flash floods, mudslides and debris flow in Ladakh. It is a cold desert which receives only 15mm rainfall in August. On that unfortunate day of 6th August, 2010, the area is said to have received rainfall of at least 150mm and 250mm. More than 71 villages of the Leh were destroyed. People were unaware as the cloudburst happened sometime in night. A lot of climate change has seen in the Jammu and Kashmir in past two decades. Summers have dry conditions like droughts receive little rainfall in installments. This has great implications for the region.

Kedarnath Floods 2013

Natural disasters are powerful forces that cause considerable loss of life and property. Human life, economic infrastructure, and natural resources in and around the affected areas are severely impacted. More number of people will be relocated and are should struggle to re-establish their homes and recover their livelihoods. Kedarnath is one of the oldest and most important worshipping place for the people of Hindu religion located in Uttarakhand state of India. It is situated in the snow cover region of Himalayas at the height of 3,583 meters approximately above sea level in the Mandakini valley of Rudraprayag District, Uttarakhand. This area is seismically and ecologically very susceptible and subtle even a minute change in terms of anthropogenic or natural can produce a dangerous catastrophe. The brittle nature of oldest crystalline basement of the Himalayan is extremely sensitive in case of landslides and any catastrophe. The Rudraprayag district where

Kedarnath is located has previously faced the trouble of natural disasters for about eight times for last 34 years. During 2012, Okhimath area of Rudraprayag has also met unprecedented damage to the life and property, infrastructure and landscape during 13th to 16th of September due to pouring rainfall and cloud burst accident. The complete area, which accommodated around 100 to 150 shops and five hotels, to serve the needs of the pilgrims, was wholly washed away leaving no trace of them.

The excessive loss of life and obliteration of assets and livelihood has considerably raised public wakefulness among the people in Uttarakhand and other people living in similar geographic locations around the world about the defenselessness of natural and human environments to natural disasters. These impacts also created a demand, especially among the people located in hill stations in India, for more information about how natural disaster affect the tourism sector which is said to be a major contributor to Indian GDP and how best to look after recovery.

Fig. 28: In this photograph taken on June 20, 2016, an Indian Hindu devotee pays his respects at the rock which partially diverted floodwaters around the Kedarnath Temple (R) during the 2013 North India floods, in the Rudrapryag District of northern Uttarakhand state.|AFP (Indian Express 26 april 2018)

Maline Landslide in Pune

Maline Landslide occurred in Pune, more than 100 people were killed. The tragedy occurred on 30th July 2014 when an incident of landslide suddenly wiped out village of Maline. This was the worst landslide of the year. There are two types of landslides- naturally induced and human induced. The current landside was possibly due to human induced activities like road construction and farming as mentioned by the geologists. Reports suggested that due to huge construction, deforestation and farming activity the landslide was occur. Scientist were warned to the in the past when some smaller landslips had occurred.The flow of the backwaters of the Dimbha Dam was also one of the causes. But no preventive measure was taken by the government. JCB machine were being used on the hillside for developing the Adivasi lands. The use of these heavy machines further increases the risk and damage the hillsides. The Adivasis were strongly opposed the use of machines, but the government did not stop their use and as a consequence landslide occurred.

Credit : IB Times Staff Reporter

Fig. 29: Maline Land Slide

Kashmir floods 2014

Heavy rain of four days caused severe flood in Kashmir valley in 2014, this is also caused by some anthropogenic activities, and therefore it was looked at as a man induced disaster. The state of Jammu and Kashmir witnessed its worst flood in 60 years. Over 0.6 million people were displaced and about 200 were died. According to the reports disaster occurred due to increasing developmental activities and a severe loss of wetland habitat for various commercial activities. According to reports the Dal Lake has reduced to just about 1200 hectares, which is almost half of its earlier spread. Chandra Bhushan, Deputy Director General of the Centre for Science and Environment quoted that,

> *"The Kashmir floods are a grim reminder that the climate change is now hitting India harder. In the last 10 years, several extreme rainfall events have rocked the country, and this is the latest calamity in the series."*

Mumbai floods of 2005, Leh cloudburst of 2010, Uttarakhand floods of 2014, followed by Kashmir floods in 2014 all are events which have more or less similar causes. According to IPCC Assessment report (AR5), floods and droughts will increase in India in future. India will get more rainfall rainy days. By 2071-2100, intense heavy precipitation over most regions will increase amount of discharge will be more in.

The south Indian Flood 2015

Heavy rain along the Caromandal coast by the north east Monsoon in November and December 2015 caused flooding in Tamil Nadu and Andhra Pradesh. The city of Chennai was severely hit by the flood, about 500 people were killed and 1.8million were displaced with an estimated damages of over US$3 billion (Rs.200 billion). The region received excessive amount of rain over 90 per cent than the amount produced by the regular monsoon, as 2015 was the El Nino year this may be the reason for this excessive rainfall. Flooding in Chennai was worsened by illegal constructions and poor flood preparedness, this flooding was described as the worst in the century. Thousands of people were evacuated from the inundated areas to safer places. The Chennai airport remained closed for five days because of inundation.

The cause of flood was human intervention also as the biggest marsh land of the city called Palikarni marshland was known as the flood sink area of the city it drained about 250 sq.km area of the city two decades ago. Due to the rapid urbanization process now these marshland is disappeared. Other wetlands were converted into waste disposal sites. The palikarni marshland became fragmented and the government found it most suitable area for urban development a survey in 2002 shows that the marsh has lost around 90 per cent of its extent. The severely affected areas during floods were lying around the marsh, after the disaster people started the movement to save the marsh land.

Fig. 30: Tamil Nadu floods: People being rescued after heavy rains caused floods in Chennai. (Photo: PTI)

Funding Arrangements of disaster management activities

Government at the central as well as state levels have some schemes for funding of disaster management activities be it relief, rehabilitation or reconstruction. The Calamity Relief Fund (CRF) is one such fund at the central level. The main responsibilities of CRF are disbursement of relief food.

The CRF is a centrally sponsored scheme of the government of India. The central government plan and approve all the centrally sponsored schemes. Ministry of Home affairs is the nodal authority for this scheme every year government contributes 75 per cent of the funds for a state and the rest 25 per cent comes from the state governments.

The Main objectives of CRF scheme are:

- Enable states to expand on the calamity relief
- Avoid delays in a state government's response to the damage caused by a natural calamity.
- Discourage the states from inflating their demands for the relief funds.
- Provide greater autonomy and responsibility to the states in relief operations.
- Make the states more accountable in the area of relief operations in post disaster period.

The CRF is limited to only six natural disasters. There is a need to increase the quantum of allocation for relief work as well as reconstruction and rehabilitation.

National Fund for calamity Relief

The Tenth Finance commission considered the issue of a calamity of rare severity, this fund is managed by a National Calamity Relief Committee, which would have representatives from the centre as well as the states. The committee is headed by the Union Minister of Agriculture comprising of Deputy Chairman of Planning Commission.

Assessment of damage could be undertaken by an independent body of experts. For this purpose, it recommended the establishment of a National Centre for Calamity management (NCCM) for monitoring the natural calamities.

MPLADS

A funding arrangement called the members of parliament Local Area Development Scheme (MPLADS) was started in December 1993. Under the scheme, MPs are allotted funds annually to pursue development works in their constituencies. Implementing agencies can be the government bodies or Panvhyati Raj Institutions (PRIs) or any other reputed NGOs that are capable of implementing these activities satisfactorily. The scheme has a lot of potential, funds of Rs.2 crores are earmarked per constituency.

The Prime Minister's National Relief Fund

The Prime Minister's National Relief Fund, created after independence, and provides immediate relief to people in distress. The fund depend entirely on voluntary donations received from the public. The fund renders assistance to individuals facing disaster situations. It is utilized to provide immediate relief to the disaster victims and the families of those killed in flood, cyclones, and earthquakes. It also provide assistance to people affected by riots, and accidents.

Insurance schemes

Disaster insurance can also be possible through a system of micro-finance. Many Micro- finance Organizations are now working in disaster-prone areas. The success of MFOs to coordinate relief organizations depending on the terms and conditions of the loans. The insurance agency may promote a community rating system to encourage communities to go beyond the required standards of minimum safety. The incentives could be a reduction in insurance premium for policyholders within communities that take appropriate actions to reduce disaster losses.

Other schemes are Swrnajayanti gram swarozgar yojna, National agricultural insurance scheme etc.

Seismic Vulnerability of National Capital Region Delhi

The national capital region lies in seismic Zone IV as per Indian Seismic Code. Delhi had experienced local earthquakes with epicenters within about fifty kilometers radius. The magnitudes of such near earthquakes are up to 5.75 to 6.0 on Richter scale. In addition, Delhi has been experiencing seismic vibrations due to large magnitude (M >7.5) in the Northwestern Himalayan Region and also the Hindu Kush region. The damages by local earthquakes are exclusively by Shear Waves. The damages due to distant earthquake in Himalayas are due to Rayleigh Waves (also known as Long Period Seismic waves). These two types of damages are due to entirely different waves. The damage due to Rayleigh Waves is only to tall structures of height more than 17 meters and is confined to the distance 150 to 550 km range from the epicenter. Such damages were seen during Bhuj earthquake at Ahmedabad. The city of Ahmedabad is located at a distance of about 320 km from the Bhuj earthquake epicenter. At Ahmedabad only tall structures have suffered heavy damage while buildings up to two or three floors have not suffered any damage or the damage was minimal. Similar situation could become a reality if an earthquake of magnitude 7.5 or more occurs in Himachal or Uttarakhand. Incidentally the IS Code does not provide any requirement for protection from Rayleigh Waves to the tall buildings. At present the probability of occurrence of a large earthquake in Uttarakhand is as high as 0.98. If such a seismic contingency occurs the damage scenario at Delhi would be a gory picture.

The Seismic Waves

The National Capital Region (NCR) with the present population of about fourteen Millions is exposed to seismic disaster. The damages are due to two types of waves. The first one is due to local earthquakes, which may occur within about 30 to 50 km radius from Delhi. Such damages are due to S Waves. The other type seismic damage could be due to the occurrence of large magnitude (magnitude more than 6.5 to 7.0) earthquake in the Himalayas. The damages caused by these two earthquake waves are entirely different. As per IS code 1893, the National Capital region (NCR) of Delhi lies in Seismic Zone IV. The basic horizontal seismic coefficient as recommended by the IS Code 1893, is 0.05. The buildings are (supposed?) designed as per the IS Code 1893. Historical records tell about earthquakes experienced at

Delhi. A number of authors such as Bilham (1995), Bilham and Gaur (2000), Khatri (1999), Seeber and Armbruster (1981) Bapat et al (1983) have discussed the seismicity of Delhi. The NCR region has been studied for seismicity and a number of Micro zonation studies have been made. Examination of past historical seismic data indicates that the NCR region of Delhi could experience a medium to moderate size earthquake of magnitude less than or equal to 6.0 on Richter scale. The latest recorded such earthquake had occurred in 1960 and details about it are as given below.

Table 19: Significant earthquakes in and around Delhi since 1720

Date	*Lat (N)*	*Long (E)*	*Mag*	*Region*	*Dist from Delhi*	*Epcl Int*	*Delhi MM*
1720 07 15	28.37	77.10	6.5	Delhi	27 km	X	X
1803 09 01	27.50	77.70	6.8	Mathura	129 km	IX	VII
1842 01 16	27.00	78.00	6.0	Nr. Mathura	192 km	--	--
1956 1010	28.15	77.67	6.7	Nr. Bulandshahar	65 km	VIII	VII
1960 08 27	28.20	77.40	6.0	Nr. Faridabad	46 km	VII	VII
1966 08 15	28.67	78.93	5.8	Nr. Moradabad	167 km	--	--

This earthquake was surveyed and reported by Mukti Nath et al (1960) of Geological Survey of India (GSI). The top portion (a minaret type mortar pole of about two meters length) of clock tower of Rashtrapati Bhavan was thrown during this earthquake. The damage to various Government buildings was estimated to be of about Rs. 2 to 3 Lakhs. This is shown in Photo 1(a) and (b)

Date:......... 27 August 1960

Time :....... .21 28 00 (IST)

Location..... 28.6 N and 76.7 E

Magnitude: . 5.75

Intensity:......VII

The epicentral track of the earthquake was between Palam to Gurgaon. This description is for local earthquake. During recent years, two small magnitude earthquakes were recorded and felt in Delhi. Bansal et al (2008) have discussed these events. The notable significant seismic events in and around Delhi are given in Table 1. (as reported by Bansal et al (2009).

Statistical Prediction of Earthquake by GSI: The Geological Survey of India (GSI) after three years of field studies has observed that the probability of occurrence of a large magnitude earthquake (M > 7.5) in Uttarakhand is in the range 0.86 to 0.98. This means that the earthquake is just at the door. It could occur any time. Bapat (2007) has discussed this probabilistic prediction. Though this observation is for Uttarakhand it could equally affect Himachal Pradesh. In a broad sense it could be said that the predicted earthquake could occur in the conglomerate of Himachal and Uttarakhand. If such an event occurs in any one of the state then the possible intensity in the epicentral region would be between IX to X on MM Scale and an area of about 50 to 70 km radius would suffer heavy damage. The other regions would also be affected

Various Types of Seismic Waves: The seismic waves have been divided in four major classes. The first is P wave which has a push and pull type motion. The second type is S Wave, which is also known as shear wave, and it has up and down type motion. The third one is Love Wave and the motion in these waves is sideways in horizontal plane. The last one is Rayleigh Wave or Long Period Surface Waves are the slowest waves from an earthquake. The motion in these waves is elliptical in vertical plane. These waves travel slowly and have very long period of the order of 12 to 18 seconds. The destruction in the epicentral area is mostly due to S Waves. The IS Code provides suitable guidelines for design and construction of buildings to protect the structures from adverse effects of S Waves. After the Bhuj earthquake and damage at Ahmedabad, the IS Code was not revised or amended for tall structures and the likely effect of Rayleigh Waves.

The Rayleigh waves have a peculiar characteristic. During the distance strip between 150 km to 550 km or so, the amplitude of these waves are higher than at other distances. Considering the long period of the waves, such waves cause vibrations in tall structures of height more than 17 meters. The vibrations induced by Rayleigh Waves cause heavy damage to tall structures situated within the vulnerable distance strip of about 400 km length.

Damages from Rayleigh Waves: During the Mexican earthquake of 1985, the tall structures in Mexico City have suffered very heavy damage. The epicenter of this earthquake on 19 September 1985 was located at a distance of about 530 km from Mexico City on Pacific coast. The magnitude of earthquake was 8.1 on Richter scale. The level of acceleration as recorded by various instruments was of the order of about 0.17 g in Mexico City. Tall buildings in Mexico City suffered heavy damage while houses with one two or three floors did not suffer any damage or the damage was minimal. Singh (1988) has discussed this earthquake in details.

During the Bhuj earthquake on 26 Jan 2001, the tall structures in Ahmedabad City have suffered heavy damage. Ahmedabad is situated at a distance of about 320 km from the epicenter of the Bhuj earthquake. In Ahmedabad two / three storey houses have not suffered any damage or the damage was minimal. There is one more example from Islamabad in Pakistan. The 8 October 2005 earthquake of magnitude 8.0 had destroyed the only high rise buildings in Islamabad located at a distance of about 150 km from the epicenter. These three examples from recent past are enough to give idea about the effect of Rayleigh Waves on tall structures located within a distance of about 250 km from epicenter.

Historical evidence of damage of tall structure at Delhi: There was strong earthquake in the Himalayan region on 01 September 1803. Oldham (1893) has described this earthquake. The upper two floors of the 71 meters tall Qutab Minar were thrown out during the seismic vibrations. A stone plaque on the inner walls of staircase of Qutab Minar mentions about the damage and subsequent repairs of the Minar. From modern structural designs the Qutab Minar is just a brick and mortar structure with a tapering cross section. There are no reinforcements in the Minar. A tall building with a tapering cross section helps in keeping the center of gravity at an height less than h / 3 from ground, where h is the height. If the center of gravity is at a distance more than h / 3, which could be a case if the tall building has uniform cross section, then the possibility of damage is very high.

Tall structures in NCR of Delhi

The population and number of tall structures in Delhi have increased almost in geometric proportion since mid-eighties or so.

The construction of tall buildings begun sometime in middle of eight decade of last century. During 1990 the number of tall buildings have increased. Further, the tall structures are somewhat concentrated in few localities such as Connaught Place, Nehru Place etc. There are more than 100 tall buildings in Delhi municipal area and an equally significant number of buildings in nearby areas of Noida, Gurgaon, Faridabad and Ghaziabad. Most of the buildings have uniform cross section for all floors. An earthquake of large magnitude (more than 7.5 on Richter scale) would cause heavy damage within a radius of about 60 to 80 km from the epicenter. In areas between 80 to 120 km radius there could be non-repairable damages and some repairable damages. Beyond this distance the damage would depend on different factors such as geology, type of construction, age of the building etc. But Rayleigh Waves from such an earthquake would definitely cause heavy damage to tall structures in NCR Region.

There are several road bridges, rail bridges, and fly-over in Delhi. There are a few multi-storey buildings having floating columns and projected cantilevered balconies. Such buildings are highly vulnerable. Another disturbing factor is structures on stilted floors with ground floor earmarked for parking. In Ahmedabad, such buildings, up to three or four floors have suffered heavy damage. Most of the Micro zonation exercises in Delhi have clearly pointed out that the trans-Yamuna localities, could suffer more damage or the intensity of damage would be higher. But most of Micro zonation distributes the area as per geological considerations. For proper appraisal of damage, the area needs to be grouped as per the level of acceleration. Singh et al (2002) have discussed this point. Delhi has another peculiar characteristics. Due to shortage of space, lands have been reclaimed for construction and at some locations the buildings are standing on filled land. There is a significant observation from Kobe (Japan) earthquake. It was seen that about 95 % earthquake damage was concentrated on filled lands. Sometime the material on reclaimed land could be of uniform geological nature. But in the case of filled land the material would heterogeneous and it could enhance the damage.

There are several bridges over Yamuna River. Geologically Delhi is situated on soft ground except some small patches of hard rock. It is seen that one side of the bridge on Delhi side has somewhat good quality of ground. While on the other hand the side in trans-Yamuna localities, the bridges are resting on very soft clay/soil. Under seismic vibrations it is possible that these bridges may suffer some damage. The national and International airports are very busy. It is extremely essential that the Airport Authority and Staff of various Airlines should undergo seismic drill and suitable training.

Seismic Disaster Mitigation and Management

At present, there is no administrative mechanism to warn the people about imminent earthquake. As such people are caught unaware and this causes more damage. The present seismic disaster mitigation and management plans are mostly for post-seismic periods. These are based on archaic principle that 'earthquake cannot be predicted'. This is mostly clearing of debris and corpses. A disaster mitigation and management plan, which cannot save, lives of people needs to be changed and overhauled. As such by its nature it deals with rescue, evacuation, rehabilitation, medical aid, food, tents, blankets, mobile X-Ray machines, jeeps, earth moving equipment etc. The present disaster management plans are not adequate to save lives of large population because there is no mechanism of issuing a warning. Most of the disaster managers feel that earthquakes cannot be predicted. They are right to some extent. Till now there has been only one case of successful earthquake prediction in China. On the other hand a number of researchers are coming forward with enough evidences and that too huge level of confidence that earthquakes can be predicted. 2008 has admitted that during the Sichuan earthquake there was no proper co-ordination between Administration and Researchers. Had there been a proper and cohesive relationship the losses could have been reduced considerably. Incidentally, the social, geological, seismological situations in China and India are comparable to some extent. It would be in the interest of saving lives of several people that a suitable approach is developed for earthquake disaster mitigation for the Northwest Himalayan situation. Even if the Administration does not issue any warning, the public education and creation of seismic awareness should be so powerful that people would draw their own conclusions.

Incidentally some of the owners and builders may argue that tall buildings in NCR Delhi have been constructed as per IS Code. This is correct and all right for epicentral damage. But this is not correct for distant damage due to Rayleigh Waves. Mexico has revised their seismic code after the 1985 earthquake damage. But the Bureau of Indian standards is yet to initiate any action about codal revision. As a matter of fact this should have been considered after the Bhuj earthquake in 2001. The code was revised and the major revision was reducing the number of Seismic Zones from five to four and abolishing Zone I. In seismically active countries such as America, China, Japan etc the tall structures are protected from possible damage due to Rayleigh Waves, also known as Long Period Waves.

Conclusion

It is thus seen that the potential danger of damage to tall structures in NCR of Delhi is real if a seismic event of large magnitude ($M > 7.0$ to 7.5) occurs in the Northwest Himalayan region. The possibility of a large magnitude earthquake has already been reported and accepted by a number of researchers. It is hoped that the concerned engineering and administrative agencies such a Central Public Works Department (CPWD), Bureau of Indian Standards (BIS), Central Building Research Institute (CBRI), National Institute of Disaster Management (NIDM), National Disaster Management Authority (NDMA), Delhi Disaster Management Authority (DDMA) etc should ponder over this problem seriously and initiate appropriate action. If no action is taken immediately, it is quite possible that the scenarios at Mexico City and Ahmedabad may be repeated in the NCR Region of Delhi.

Forecasting of Disasters in India

Britishers started cyclone, flood forecasting and also established an earthquake observatory in India. Cyclone monitoring station was established in 1865 at Kolkata, west Bengal, Earthquake observatory in 1898 at Kolkata in West Bengal, a flood forecasting station was started in New Delhi at Yamuna River Bridge after independence in 1958.

Britishers wanted to know the disaster regime of India because they want to know the vulnerable regions before the development of infrastructural facilities including settlements, railways, roads and construction of public buildings. Kolkata was their capital so they developed the monitoring of cyclones and earthquakes here first.

In 1864 two severe cyclones struck the eastern part of the country one at Kolkata in October and the other at Machlipatnam in November, Government quickly developed a system of cyclone warnings in 1865, later by the year 1880 other port cities of Mumbai, Karachi, Karwar and Kumta etc. By 1886 Cyclone warning systems were developed all over the Indian coast. Mumbai was the headquarter and served all the west coast ports.

Now Indian coast has dozens of cyclone warning centres and cyclone detection radars, which can track cyclones within a range of 500 km, beyond it satellites imageries are used to track the cyclones.

First flood forecasting and warning network was established in 1958 at Yamuna in Delhi. Now we have flood forecasting station almost at every river basin of the country covered by 175 flood forecasting stations.

Table 20: Number of flood forecasting sites in major inter-state river systems.

S. No.	*Major Interstate River Systems*	*Type of Forecasting Sites*		*Total*
		Level Forecasting	*Inflow Forecasting*	
1.	Ganga &its tributaries	77	10	87
2.	Brahmaputra & its tributaries	27	00	27
3.	Barak System	05	00	05
4.	Eastern Rivers	08	01	09
5.	Mahanadi	03	01	04
6.	Godavari	14	04	18
7.	Krishna	03	06	10
8.	West Flowing Rivers	09	06	15
9.	Southern River System (Pennar)	01	00	01
	Total	**147**	**28**	**175**

Source: CWC.

Table 21: State-wise Distribution of Flood Forecast Stations of Central water Commission

Sl. No.	*Name of State/UT*	*Number of flood forecasting Stations*		
		Level	*Inflow*	*Total*
1.	Andhra Pradesh	4	8	**12**
2.	Arunachal Pradesh	1	0	**1**
3.	Assam	25	0	**25**
4.	Bihar	32	0	**32**
5.	Chhattisgarh	1	0	**1**
6.	Gujarat	6	5	**11**
7.	Haryana	0	1	**1**
8.	Jammu & Kashmir	1	0	**1**
9.	Jharkhand	2	5	**7**
10.	Karnataka	1	7	**8**
11.	Madhya Pradesh	2	2	**4**
12.	Maharashtra	7	3	**10**
13.	Orissa	11	1	**12**
14.	Rajasthan	0	1	**1**
15.	Tamil Nadu	0	5	**5**
16.	Telangana	5	5	**10**
17.	Tripura	2	0	**2**
18.	Uttar Pradesh	34	2	**36**
19.	Uttarakhand	3	0	**3**
20.	West Bengal	11	3	**14**
21.	Dadra & Nagar Haveli	1	0	**1**
22.	NCT of Delhi	2	0	**2**
	Total	**151**	**48**	**199**

Source: Ministry of water Resources.

Besides there are more than 400 stations for recording of rainfall, humidity and temperature for drought monitoring done by Indian meteorological department, Central Water Commission monitors underground water by making measurements of seventy-one major reservoirs across the country. Droughts monitoring also includes estimation of water level in different types of storage.

Table 22: Vulnerable population at the Macro-Regional Level in India

Macro Region	*Area (in sq. Km)*	*Total Population*	*Vulnerable population*
Himalayas	420,779	44,460,754	25,817,378
Northern Plain	733,849	416,884,509	197,942,423
Deccan plateau	1,529,976	372,344,938	200,755262
Coastal Plains and Island	481,690	194,907,374	86,513,467
India	**3,166,294**	**1,210,854,977**	**511,028,530**

Census of India 2011.

In the Himalayan region about 58.07% of the total population of the region is vulnerable. The Deccan plateau stands seconds with 53.92 per cent, the northern plains have 47% of its total population vulnerable from disasters. The coastal plains and islands have 44.30% vulnerable population. If we make a profile of the vulnerable population of India it is evident from the above data that share of vulnerable population increases from west to east. The Himalayan region has 85% of its population as vulnerable but the eastern states it increases up to 90%. The Deccan plateau has more than 50% of its population is vulnerable from disasters. In the Ganges plain Vulnerability varies there are some pockets of least vulnerability in the Ravi –Beas region, but in the lower Ganga plain the vulnerability is higher than the national average. It further increases in the Barind Tract and Rahr basinwhere vulnerable population is nearly 80%.

Table 23: Number of Observatory Stations for forecasting the natural hazards

Years	**Seismic observatories**	**Flood Forcasting stations**	**Meteorological stations**
Prior to 1900	1	0	119
1901-50	4	0	233
1951-60	10	1	346
1961-70	18	43	390
1971-80	26	84	391
1981-90	40	157	391
1991-2000	49	243	391
2001-03	66	243	391

Based on unpublished records of the Indian meteorological Department, and CWC.

Key Concepts

AERB	:	Atomic Energy Regulatory Board.
CMG	:	Crisis Management Group.
DAE	:	Department of Atomic Energy, Government of India.
NCMC	:	National Crisis Management Committee.
Composite Disaster Mapping	:	Disaster maps which indicate the occurrence and Impact of a number of different disasters over the same area.
Carbon Sink	:	Trees absorb Carbon di oxide and build up the same in The form of wood. Thus they act as carbon sinks.
Chemical pollution	:	The effluent which is not organically degradable, or Cannot be treated by a biological process.
Crown Fires	:	The most unpredictable fires, they burn the top of Trees and are spread rapidly by winds, very difficult to control until it drop on the ground.
Disaster Mapping	:	Disaster mapping is a tool for assessing, storing and conveying Information the geographical location and spread of the effects of Disasters.
Desert Development programme	:	This was launched in 1977 on the recommendation Of the National Commission of Agriculture. The aim is To control the

desertification process and raise the Productivity of these areas.

Effluent : Wastewater generated in the manufacturing process Of Industrial products.

Equalisation : Storing the liquid wastes for a certain period of time and mixing in a holding tank for reducing the fluctuation

Eutrification : When Phosphorus and nitrates from industrial Effluents are discharged in a water body, they act as Nutrients and promote the growth of Oxygen, consuming algae, thereby reduce the level of oxygen of water, kill fish and other aquatic organisms.

Mercalli Scale : Earthquakes are generally measured by their intensity and magnitude. Its intensity differs in appropriate locations, depending upon the distance of the location from the epicenter. The ranking system used to measure seismic activity.

Micro zonation : The identification of separate individual areas having different potentials for hazardous earthquake effects.

Lahars : Heavy rains or melting snow can turn the Pyroclastic Material on steep slopes of a volcano into mud and Set off a catastrophic mudflow called Lahar.

Organic load :

Pandemic : Large epidemic that spread worldwide or in a large region.

Reconnaissance : It is the military observation of an area to gain Information.

Segregation : Strong and weak wastes, or waste of different Characteristics may be separately collected, Treated and disposed.

Shake Map : It is a representation of ground shaking produced by the earthquake. Shake maps indicate the ground motions.

Shear : Cracking and breaking of the rock by compression Or tension at a fault or thrust plane.

Sulfur Mustards : Sulfur mustards are vesicants and alkylating agents In vapour and liquid forms can be absorbed through The eyes, skin and mucous membranes.

Watershed Development project : The programme implemented since 1993 aims at

In Shifting cultivation areas : development of Jhum areas, reclaiming the jhum

Lands and encourage the Jhumias for settled agriculture.

Zoning : It is the earmarking an area of land or sea from the point of occurrence of the disaster.

References

Aandahl, G. and K.O. Brien.2002. Damage, Seissmotectonics and Isoseismals for the Kutch Earthquake of 26th January, 2001.New York: Ellis Harwood Limited.

ADPC (2002).Community Based Disaster Management: Trainers Guide, Asian Disaster Preparedness Centre, Bangkok, Thailand.

Agrawal, P.N.(2000). Seismological aspects of earthquake damage reduction.Sixth IGC foundation Lecture, IGC-2002 Udaipur.

Alexander. (1993). Natural Disaster. UCL Press, London.

Anonymous (2005). Tsunami disaster-Relief and rescue operations. Employment News, 29(43):56

Anonymous (2004)Integrated water resource Management. Dublin Conference, WMO, Geneva, Switzerland, p.1.

Anonymous (2008). Times Of India. 29th August, 2008.

Ariyabandu, Malagoda, Madhavi, 1999, Defeating Disasters: Ideas for Action, A Duryog Nivaran Publications, Sri Lanka.

Asian Disaster Reduction Centre. Top 25Natural Disasters in India according to Amount of Damage(1901-2000) Available online at Http//www.adrc.asia/databook-20th/INDI.XLS

Bana, S.2001. Devastation in Gujarat, Business India, 598: 50-58.

Bapat.A.R.C, Kulkarni and S.K.Guha.1983: Catalogue of Earthquakes in India and Neighborhood from Historical period up to 1979.Roorkee.India Society of Earthquake Technology. Indian Institute of Technology.

Barton, Allen, 1970, Communities in Disaster : A sociological Analysis of Collective Stress Situation, Doubleday Anchor Books, New York.

Basu, A.R. and R.Srinivas Murthy.2003: Disaster and Mental Health: Revisiting Bhopal, Economic and Political Weekly, 38(11), 1074-82.

Bell, F.G.(1980). Engineering Geology and Geotechniques. Newnes-Butterworth, London.

Bhatt, C.P. and N.Juyal.2000.'Living with Earthquake: Lessons from Chamoli and Rudraprayag(Central Himalayas), Science and culture, 66(1-2): 16-19.

Bhargava.P.M1985, The Bhopal Tragedy: A Middle World, Economic and Political weekly, 20:962-65.

Blaiki, P.M. Canon, T.Davis, I. and Wisner, B.(Eds.)[1994].At Risks: Natural Hazards, peoples vulnerability and disaster. New York, USA.

Blanford, H.F.1877, Catalogue of the Recorded Cyclones in Bay of Bengal up to the end of 1876, Journal of the Asiatic Society of Bengal, 46 (2): 328-38.

Bose. A.2000. Are Natural Disasters Manmade? Economic and Political Weekly, 35 (43-44)

Bose, S.and saxena, N. (2000). Tropical Cyclones. Employment News, 25(4):1-3.

Bruce, A. (1978) Earthquakes: A Primer. W.H. Freeman and company, USA, P.240.

BUCKLAND, jerry and M. Matiur Rahman, 1999, Community Based Disaster Management during the 1997 Red River Flood in Canada, journal, Disasters.

Carter, N.(1991). Disaster Management: A disaster Manager's Handbook. Asian Department Bank, Manila, Philippines. Central Water Commission.1985, Flood Atlas of India. Delhi: Government.

Chakraborti, A.K. Bhaunamurthy, V. and Chandrasekhar, G.(2003). Remote Sensing towards filling the gaps in flood monitoring impact assessment and forecast modelling. In: Bulletin of the National Natural Resources Management System, NNRMS (B)-28(Eds. V. Jayaraman, R.R. Navalgund, S.K.Bhan, S.K. Shivkumar). Department of space, Government of India,

Bangalore, pp.44-51as.K.2002Social mobilization Relief Work in Cyclone affected Orissa, Economic and political Weekly, 37(48) 4784-788.

Clarke, Tim (1991). EIA and Sustainable Development. EIA Trainers Newsletter, 6 central University of Manchester, U.K.

Coburn, A.W, R.J.S. Ssppence and A.Pomonis, 1994, Vulnerability and Risk Assessment, Disaster Management training program, United Nations.

David Day, M. M. Harrison and S. M. Halpin, (2009)An Integrative Approach to Leader Development: connecting adult development, identity, and expertise, New York: Routledge.

Ganesh Raj, K. and Nijagunnappa, R.(2004).Major lineaments of Karnataka state and their relation to Seismicity: A remote Sensing based analysis. Geological Soc. India, 63: 430-439.

Gaur, V.K.(2003). Mitigating adverse impacts of natural hazards: The crucial role of space technologies. In: Bulletin of the National Natural Resource Management System, NNRMS-28(Eds.V.Jayaraman, R.R.Navalgund, S.K.Bhan, S.K. Shiv kumar). Department of space, Government of India, Bangalore, pp.12-23.

Ghani, M.U.(2003) Participatory strategy for flood mitigation in East& North-East India: A case study of Ganges- Brahmaputra-Meghan basin. Unpublished Report(http://www.unescape.org/esd/water/disaster/2001/india.doc)

Ghosh, S.N.(1997). Flood Control and Drainage Engineering. Oxford & IBH publication, New Delhi.

Giardini, D. and Basham, P.(1993). Global Seismic Hazard Assessment.

Glade, T. and Crozier, M.J (2005). Landslide Hazard and Risk. John Wiley & Sons, New York.

GOI (2004). Disaster Management in India: A Status Report. NDM Division, Ministry of Home Affairs, Govt. Of India, New Delhi.pp.1-86.

Goi(2005) Disaster Risk Reduction: The Indian Model. NDM Division, Ministry of Home Affairs, Govt. Of India, New Delhi.

Green.T.H. and T.H. Gross9eds).1992.The Philosophical works of David Hume(1882-86)Aalen:Scietific Verlage.

Gupta, M.C. (Ed) 2001.Natural Disaster Management in India. New Delhi. National Centre for Disaster Management, Indian Institute of Public Administration.

Gupta, M.C., L.C. Gupta, B.K.Tamini and Vinod. K. Sharma (eds.) 2001.Manual on Natural Disaster Management in India. New Delhi National Centre for Disaster Management, Government of India.

Gupta. M.C. and V.K. Sharma.1999.orissa: Super Cyclone 99. New Delhi: National Centre for Disaster Management, Indian Institute of public Administration.

Holland, Gj.(Ed.)(1993) Global Guide to Tropical Cyclone Forecasting. World Meteorological Organization, WMO/ TD.No.560, Report No.TCP-31.

Hooker, j.(2008). Toll Rises in China Quake. New York Times (Retrieved on 2008-06-01.

http://www.nature.com/nature/debates/earthquake/equake.7.html

http://origin.www.spectrum.ieee.org/login

http://www.spectrum.ieee.org/login

IPCC, 2001, Impacts Adaptation and Vulnerability, GRID, Aewndal.

Jai Krishna (1992). Seismic zoning maps of India. In: Bulletin of the National Natural Resources Management System, NNRMS (B)-28(Eds.V.Jayaraman, R.R.Navalgund, S.K.Bhan, S.K. Shivkumar). Department of space, Government of India, Bangalore, 17-23.

Joseph, P.V.(1994). Tropical Cyclone Hazard and Warning and Disaster Mitigation systems in India. Sadhana Academy proc. Engineering Science. Indian Academy of Science, Banglore, 1955-56.

Kapur.A.(2010).Vulnerable India: Sage publication, USA,

KOpomaa, T, 1999, Constructing and preparing the City for Crisis: Before and After: Review of some Finnish and International cases, Helisinki unpublished conference paper.

Kulshreshtra, S.M.(1997). Drought Management in India. Joint COLA/ CARE Technical Report No.1. Institute of Global Environment and Society, USA.

Kumar, S. and Parmar, A.(2005). Vulnerability and Hazards reduction strategies in India. In: Environment & Development: Challenges and Opportunities (Ed. Jagbir Singh). I.K. International publishing House Pvt. Ltd., New Delhi, pp.375-392.

Mahlawat, Manjeet and Singh, Mahadevi(2003). Impact of drought and its management. Chapter3, In: Current Environmental Issues (Eds.B.B.s. Kapoor, Ahmed Ali, K.K.Singh and Chandrakanta). Madhu Publication, Bikaner, p.37-52.

Mushir.S .M. & Khan.M.F. (2007). Quality Of Urban Environment: Idaraye Adabiyat –i-Dilli, Daryagunj, Delhi 11007.

Russell, G.(2005). The humanitarian relief supply chain- Analysis of the 2004 South-east Asia earthquake Tsunami. Jarrod Goentzal, Ph.D. Massachusetts Institute of Technology, Cambridge, USA.

Robert G. Lord and Rosalie J. Hall, 'Identity, deep structure and the development of leadership skill', The Leadership Quarterly, Issue 16, 2005, pp. 591-615. Steven M. Norman, Bruce J. Avolio and Fred Luthans, 'The impact of positivity and transparency on trust in leaders and their perceived effectiveness', The Leadership Quarterly, Issue 21, 2010, pp. 350-64.

Sampath, P.1991, "Vulnerability Reduction at Community Level: The New Global Paradigm," in the proceedings of International Conference on Disaster Management: Cooperative Networking in South Asia, Vol.II, New Delhi.

Singh, Parbin(2001). Engineering & General Geology. 6th edn., S.K. Kataria & Sons, Delhi, pp.537-564.

Sinha, P.C.(1998)[Ed]. Introduction to Disaster Management. In: Encyclopedia of Disaster Management, Vol.1.Anmol Publications Pvt. Ltd., New Delhi.

Srivastava, H.N.(1983). Forecasting Earthquakes. National Book Trust, India, New Delhi, pp1-147

Strahler, Arthur. N. (1981) Physical Geology. Harper and Row Publishers, New York, pp.183-305.

Smith, k.(1996). Environmental Hazards. Rutledge, London, U.K.

UN (1997). Floods: People at Risk, Strategies for Prevention. Department of Humanitarian Affairs, Geneva, United Nations.

Vaidya, K.S.(1987). Environmental geology: Indian Context. Tata McGraw-Hill, Publishing Company, New Delhi.

UNDP Report on "Disaster Scenario in India" read online athttp://www.undp.org.

World Bank (200). Managing Economic Crisis and Natural Disasters. In: World Development Report2000/2001.Washington, DC, and USA.

World Disaster Report (1998) International Federation of Red Cross and Red Crescent Societies, OUP.

World Disaster Report (2016). Disaster Data. CRED, University of Louvain, Belgium.

David Day, M. M. Harrison and S. M. Halpin, An Integrative Approach to Leader Development: connecting adult development, identity, and expertise, New York: Rutledge, 2009.

S. Alexander Haslam, Stephen Reicher and Michael Platow, The New Psychology of Leadership: identity, influence and power, Psychology Press: Hove UK, 2011. Robert G. Lord and Rosalie J. Hall, 'Identity, deep structure and the development of leadership skill', The Leadership Quarterly, Issue 16, 2005, pp. 591-615. Steven M. Norman, Bruce J. Avolio and Fred Luthans, 'The impact of positivity and transparency on trust in leaders and their perceived effectiveness', The Leadership Quarterly, Issue 21, 2010, pp. 350-64.

Murlidharan, S. and S. Bukhari.1996. Death on the Amaranth Yatra, Frontline, September 20, 29: 130-32.

Murthy, R.S. 2000. Disaster and Mental Health: Responses of Mental Health Professionals, Indian Journal of Social Work, 61(4), 675-92.

Myers. (ed). 2001. Encyclopedia of Physical Science and Technology. California: Ramtech Limited.

Nandy, D.R. 2000. Plate Tectonics, earthquake and Earthquake Hazards, Science and Culture, 66.

Index

D

O

P

Q

R

S

T

U

V

W